SpringerBriefs in GIS

AF183604

SpringerBriefs in GIS present compact, concise summaries of cutting-edge research, practical applications and visualizations in the use of geographical information systems. At 50 to 125 pages (approx. 20,000 – 50,000) words, SpringerBriefs in GIS provides researchers and practitioners with an innovative venue to present work that might be longer and more complex in scope than a journal article.

This series aims to cover a wide range of topics related to geographical information science and geographical information systems. Potential topics could include: an in-depth case study on the use of GIS to accomplish a specific goal; a guide to an emerging GIS tool, technique or map; a "hot-take" on or snapshot of a current issue that needs to be published as quickly as possible (just 8–12 weeks after a manuscript's delivery and acceptance). Multidisciplinary studies are particularly welcome.

SpringerBriefs are distributed through the same channels as Springer's book content, and are available as physical books and full and chapter-wise eBooks. Both solicited and unsolicited manuscripts are considered for publication in this series. Please send questions and proposals to Zachary Romano, Associate Editor, Earth Science, Environment, and Geography, at Zachary.Romano@Springer.com.

Anindita Nath · Bappaditya Koley ·
Tanupriya Choudhury

Shoreline Modification Over the Indian Sundarban

A Geo-informatics and Machine Learning Approach

Springer

Anindita Nath
Department of Geography
Bankim Sardar College
South 24 Parganas, West Bengal, India

Bappaditya Koley
Department of Geography
Bankim Sardar College
South 24 Parganas, West Bengal, India

Tanupriya Choudhury
School of Computer Sciences
University of Petroleum and Energy Studies (UPES)
Dehradun, Uttarakhand, India

ISSN 2367-010X ISSN 2367-0118 (electronic)
SpringerBriefs in GIS
ISBN 978-3-032-13980-1 ISBN 978-3-032-13981-8 (eBook)
https://doi.org/10.1007/978-3-032-13981-8

© The Editor(s) (if applicable) and The Author(s), under exclusive license to Springer Nature Switzerland AG 2026

This work is subject to copyright. All rights are solely and exclusively licensed by the Publisher, whether the whole or part of the material is concerned, specifically the rights of translation, reprinting, reuse of illustrations, recitation, broadcasting, reproduction on microfilms or in any other physical way, and transmission or information storage and retrieval, electronic adaptation, computer software, or by similar or dissimilar methodology now known or hereafter developed.
The use of general descriptive names, registered names, trademarks, service marks, etc. in this publication does not imply, even in the absence of a specific statement, that such names are exempt from the relevant protective laws and regulations and therefore free for general use.
The publisher, the authors and the editors are safe to assume that the advice and information in this book are believed to be true and accurate at the date of publication. Neither the publisher nor the authors or the editors give a warranty, expressed or implied, with respect to the material contained herein or for any errors or omissions that may have been made. The publisher remains neutral with regard to jurisdictional claims in published maps and institutional affiliations.

This Springer imprint is published by the registered company Springer Nature Switzerland AG
The registered company address is: Gewerbestrasse 11, 6330 Cham, Switzerland

If disposing of this product, please recycle the paper.

To Lord Shiva*, the eternal source of destruction and creation—may his wisdom guide these words, and his grace dwell within every page.*

Foreword

It is my pleasure to write the foreword for this book entitled *Shoreline Modification Over the Indian Sundarban—A Geo-informatics and Machine Learning Approach.* The Indian Sundarbans, part of the world's largest delta and one of the most vulnerable coastal ecosystems, is facing mounting pressures from both natural forces and human activity. Its complex landscape—woven together by tidal rivers, dense mangrove forests, estuaries, and human settlements is under continuous threat from coastal erosion, shifting land use patterns, and sea-level rise. Understanding how this dynamic coastal zone evolves over time is essential for planning effective conservation and adaptation strategies. This author's book presents a timely and rigorous study that combines geospatial analysis with machine learning to examine shoreline change and land transformation across the region. By employing the Digital Shoreline Analysis System (DSAS), the author provides a detailed assessment of short-term and long-term shoreline modifications, drawing on multi-temporal satellite data and historical records. The DSAS methodology enables precise calculations of rates of erosion and accretion, revealing trends that are often obscured by the complexity of deltaic processes.

What sets this work apart is its integration of Support Vector Machine (SVM) classification techniques to analyze Land Use and Land Cover (LULC) dynamics. SVM, known for its high classification accuracy in remote sensing applications, is effectively used here to track changes in mangrove cover, settlements, agricultural areas, and water bodies. This dual approach allows for a holistic understanding of how natural shoreline dynamics are intertwined with human-induced land cover changes. The study's findings are of critical importance—not only to researchers and academicians but also to policymakers, environmental managers, and regional planners. As climate change accelerates and coastal hazards become more frequent and intense, this kind of integrated spatial analysis is vital for designing adaptive, evidence-based management interventions. In bridging traditional coastal geomorphology with cutting-edge remote sensing and machine learning techniques, this book offers both scientific depth and practical relevance. It provides a valuable template for coastal studies elsewhere, especially in deltaic and low-lying coastal environments facing similar pressures.

I hope the readers/researchers will find this work interesting and unique, and may get some new ideas to put forth. It is my hope that this book will contribute meaningfully to ongoing efforts in coastal resilience, conservation, and sustainable development in the Indian Sundarbans and beyond. I wish the authors all the success in getting their work published and wish them the best of luck in their future endeavors.

Arkoprovo Biswas.

Dr. Arkoprovo Biswas
Physics and Earth Science Division
Geological Studies Unit
Indian Statistical Institute
Kolkata, India

Preface

The **Indian Sundarbans**, a globally recognized ecological treasure and a vital component of the Ganga-Brahmaputra delta, stands at the frontline of climate change and anthropogenic stress. Spanning dense mangroves, tidal rivers, mudflats, and inhabited islands, this deltaic region is undergoing rapid transformations—many of which are driven by dynamic shoreline processes, sea-level rise, and shifting patterns of land use and land cover (LULC).

This book offers a comprehensive and integrated analysis of **shoreline modification** across the Indian Sundarbans, using a dual-method approach that incorporates the **Digital Shoreline Analysis System (DSAS)** for tracking short-term and long-term shoreline changes and identification of erosion and accretion zones, alongside **Support Vector Machine (SVM)** learning based classification for evaluating land use and land cover dynamics. By combining these methodologies, this study provides a deeper understanding of how physical shoreline alterations are linked with ecological and socio-economic changes across the landscape.

The **Digital Shoreline Analysis System (DSAS)**, an advanced tool developed by the U.S. Geological Survey for assessing shoreline dynamics through geospatial techniques is employed to analyze shoreline retreat and advancement over multiple temporal (2000–2020) scales—spanning both short-term and long-term (multi-decadal) periods—using high-resolution remote sensing data and historical maps providing a comprehensive understanding of the patterns, rates, and trends of shoreline migration across different sectors of the Sundarbans landscape. This analysis captures the spatial variability of erosion and accretion across critical segments of the Sundarbans coast. Concurrently, the application of SVM, a powerful machine learning technique, enables accurate classification of LULC patterns over time, allowing for the detection of changes in mangrove cover, settlement expansion, agriculture, and water bodies. Together, these analyses reveal the interconnectedness of geomorphological processes and human land-use decisions. The findings highlight regions of intense vulnerability, inform adaptive management strategies, and contribute to the growing discourse on sustainable coastal zone management in deltaic environments. This book is intended for researchers, planners, environmentalists, and decision-makers who seek a data-driven, spatially explicit understanding of coastal change in one of the most ecologically sensitive regions of the world. It also aims to contribute to the broader field of geospatial analysis for coastal research, emphasizing the importance of integrating machine learning with traditional geomorphic models.

I express heartfelt gratitude to all mentors, institutions, and field collaborators who supported this research endeavor. Special thanks go to the resilient communities of the Sundarbans, whose lived experiences provide both the context and the urgency for this work.

South 24 Parganas, India — Dr. Anindita Nath
South 24 Parganas, India — Dr. Bappaditya Koley
Dehradun, India — Prof. Tanupriya Choudhury

Competing Interests The authors have no competing interests to declare that are relevant to the content of this manuscript.

Contents

Chapter 1
Unveiling the Geomorphological, Hydrometeorological, and Geological Insights of the Indian Sundarbans

Abstract The Indian Sundarbans, being a distinctive deltaic terrain, has intricate relationships among geomorphological, hydrometeorological, and geological processes. This chapter attempts to investigate and demystify these relationships to present a holistic perspective of the dynamic geo-environmental processes of the region. Geomorphologically, the Sundarbans are influenced by deposition of sediments, tides, and fluvial interactions, resulting in perpetual modifications of the coastal morphology. Hydrometeorologically, the area is regularly hit by extreme weather conditions, such as tropical cyclones and excessive rain, heightened by the effects of climate change that affect the natural environment and human settlements. Geologically, the coastal structure of the area is regularly affected by its erosion and accretion and by changing sediment types that result in moving landforms. This multidisciplinary study brings together remote sensing information, ground measurements, and climatic models to examine the dynamically changing problems of the Sundarbans, including shoreline erosion, salinity invasion, and loss of habitat. The results provide essential inputs for sustainable management measures, building resilience, and conservation efforts, with a focus on an integrated approach to protect this ecologically rich and fragile area.

Keywords Geomorphology · Hydrometeorology · Climate change · Natural Hazards · Cyclonic Storms · Mangroves · Indian Sundarban

1.1 Characteristics of the Study Area

1.1.1 Background

Mangroves are found between 30° n and 30° s latitudes, being the major natural vegetation in the Indian Sundarbans that cover the majority of the world's tropical and subtropical regions (Ali & Rahman, 2025). They are ecologically valuable because they cover around 75% of the world's coastline (Datta et al., 2011; Anu et al., 2024). Located in the Bay of Bengal, Sundarbans takes the form of a tidal delta that

© The Author(s), under exclusive license to Springer Nature Switzerland AG 2026
A. Nath et al., *Shoreline Modification Over the Indian Sundarban*,
SpringerBriefs in GIS, https://doi.org/10.1007/978-3-032-13981-8_1

spreads both in India and Bangladesh (Ghosh & Ray, 2022). The Indian Sundarbans, approximately 9,630 km, are usually surrounded by the Dampier-Hodges Line on the northern side and the Bay of Bengal in the south. The region consists of 19 community development blocks, 13 of which are in the South 24 Parganas district and 6 blocks are in the North 24 Parganas district, both in West Bengal, in the present day. Its eastern border is Ichamati-Remgal-Hariyabanga Muhana, the Indo-Bangladesh border, and Hugli Muhana is its western border (Kar and Basu, 2023). It consists of 102 islands, out of which 54 (about 5,400 km^2) are populated (Sahana et al., 2016).

Indian Sundarbans are secured by various forest laws, i.e., the Indian Forest Act (1927), the Forest Protection Act (1980), and the CRZ Act (1991) (Kar and Basu, 2023). It is composed of conservation-core, buffer, and three regions of transition areas in terms of intensity of protection (UNESCO, 2021). It was declared a biosphere reserve in 1989, as an acknowledgment of its environmental importance. In addition, UNESCO has also classified the region as a "heritage site" due to its immense biodiversity, including diverse flora and organisms. Flat islands are characterized by a mature height due to their position near the sea level and sometimes due to floods caused by the tidal surge and monsoon. The region is inhabited by about 4.5 million people in Census, 2011. These communities rely very much on the natural resources of Sundarbans for survival, with activities such as fishing, honey collection from the Sundarbans Biosphere Reserve, and crabbing (Ghosh & Ray, 2022) serving as the backbone of their economy. Indigenous people also live on indigenous activities, such as fishing and hunting from flora and rivers, which are essential for their economic existence.

Aside from these local uses, the Sundarbans mangrove ecosystem has a critical function in the regional economy and sustainable livelihood, contributing to national economic growth through the utilization of renewable resources. The region also supports a diverse array of ecosystem services, as well as economic and cultural benefits (Barbier et al., 2011; Warren-Rhodes et al., 2011). But among the immediate challenges facing the inhabitants of the Ganga Delta is the risk of sea level rise, which might be worsened by the cumulative impacts of subsidence and climate change. On top of such challenges, the area is subjected to regular natural disasters in the form of storms, cyclones, floods, and erosion. Where fishing is the mainstay of most of its inhabitants, any interference with these waters would spell calamity for the economy of the region, considering that there are no other sources of income in this part of the region.

The Indian Sundarbans mangroves, nestled between 21°31′N 88°10′E and 22°30′N 89°51′E, are part of the largest delta on the planet, formed by the merging of the Ganges, Brahmaputra, and Meghna Rivers in the Bengal basin (Bera et al., 2022). This unique landscape is sculpted by a fascinating web of tidal channels, mudflats, and islands, all shaped by the silt that these rivers carry down from the Himalayas (Mahanta et al., 2024). To the east, the Ichamati-Raimangal River marks the boundary, while the Hooghly River lies to the west, the Bay of Bengal to the south, and the historic Dampier and Hodges lines, drawn back in 1829–1830, to the north. The Sundarbans is a low-lying coastal delta, with elevations dipping below 5 m (Bera & Maiti, 2019). Its key geomorphic features include tidal creeks, mudflats,

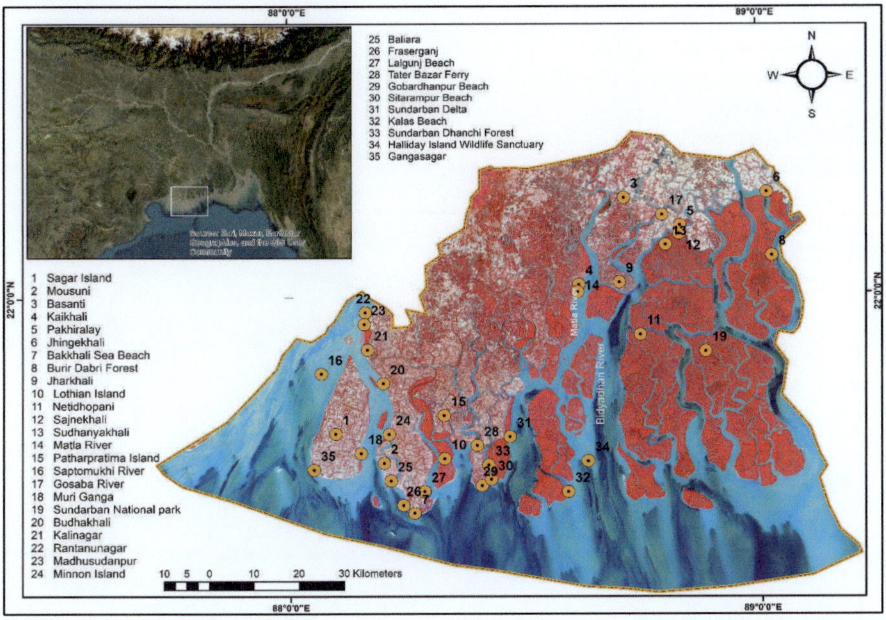

Fig. 1.1 Location of the study area

mangrove swamps, and small dunes. The region enjoys a subtropical humid climate, with average annual rainfall between 1500 and 2000 mm, 74% of which falls during the monsoon season from June to September. In June, the mean maximum temperature can reach 34 °C, while in January, it can drop to a chilly 11 °C. Cyclones are a common occurrence here, with about 3 to 4 storms sweeping through the area each year. This region is classified as meso-tidal, boasting a tidal range of 2–4 m. The study area includes 10 major islands in the Indian Sundarbans, featuring 7 from the central Sundarbans—Godkhali, Bali, Gosaba, Sajnekhali, Sudhanyakhali, Bagna, and Jharkhali—and 3 from the western Sundarbans (Bonnie Camp, Patharpratima, and Lothian Island) (Fig. 1.1). The choice of these stations was mainly influenced by the diversity of mangrove flora and their accessibility (Sreelekshmi et al., 2020). Among the 10 islands, 4 represent natural mangrove habitats: Sajnekhali, Sudhanyakhali, Bagna, and Bonnie Camp, while the others have been impacted by humans.

1.1.2 Physiographic Conditions

The study area consists of low-lying islands connected by a complex network of waterways, estuaries, creeks, mudflats, and densely forested marshy lands (Khojasteh et al., 2021; Sardar & Samadder, 2021; Sarkar, 2022). The topography features

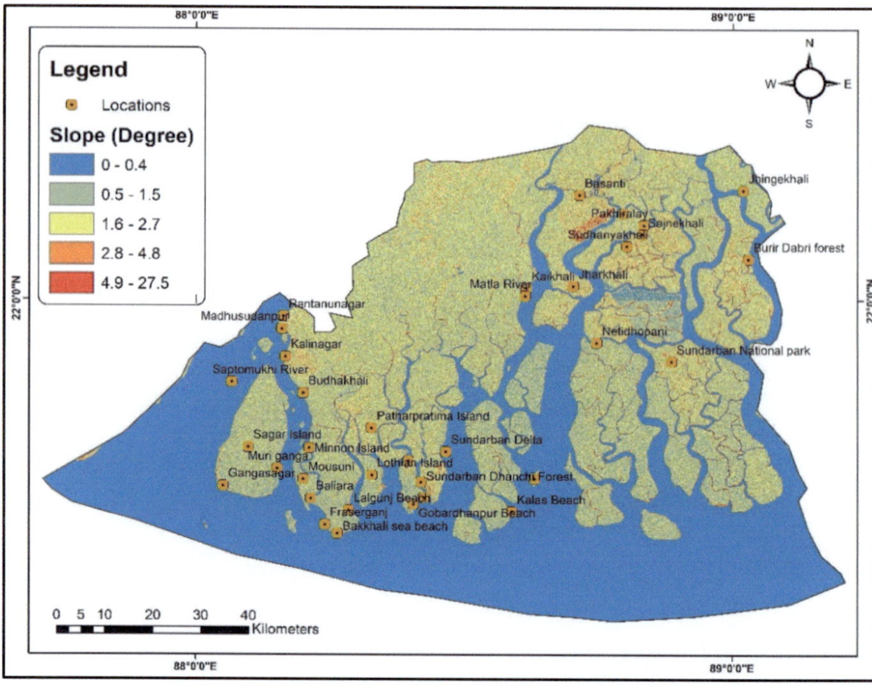

Fig. 1.2 Slope map of the study area

show gentle slopes (within 5 degrees) (Fig. 1.2) with elevations ranging up to 42 m above mean sea level (MSL) (Fig. 1.3). The deltaic regions are marked by natural levees, swamp areas, and older floodplains (Islam, 2016; Rashid et al., 2024). The present investigation shows that these areas are heavily influenced by both marine and fluvial geomorphological processes, with local rivers experiencing tidal surges (Bandyopadhyay, 2019; Bevington et al., 2022).

In the study area, we identified four distinct geomorphic units: Deltaic Plains, Levees, Marshes, and Islands. ***Deltaic Plains*** are formed by the gradual deposition of sediments from rivers and other water bodies. Primarily flat, with small hills created by sedimentary deposits from river systems like the Ganges–Brahmaputra Delta, they are characterized by alluvial soils that are ideal for agriculture, particularly rice cultivation (Islam, 2016; Rashid et al., 2024). ***Levees*** are embankments built along rivers to protect low-lying lands from flooding, particularly during the monsoon season. Without these levees, heavy rainfall combined with high tides could lead to significant flooding (Adnan et al., 2019). ***Marshlands*** are vital geomorphic features act as natural filters for pollutants, especially from nearby industries, helping to maintain the region's ecological balance (Bhuiyan et al., 2025). ***Sundarbans Islands*** provide critical habitats for numerous species, including the Royal Bengal Tiger, adding to their ecological importance (Mahmood et al., 2021). Additionally, the islands

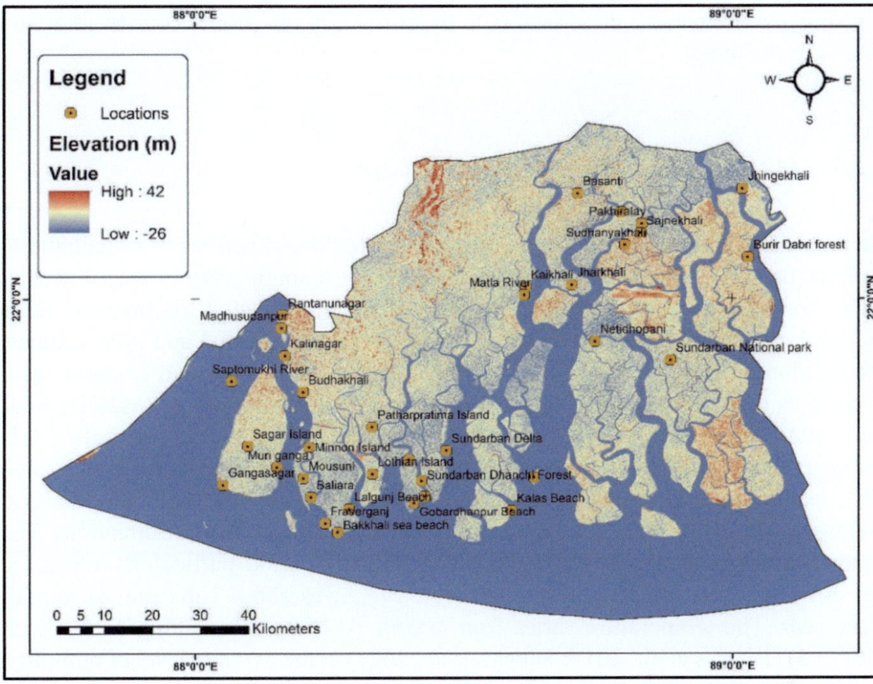

Fig. 1.3 Elevation map of the study area

support local fishing communities, offering significant livelihood opportunities for those living in the surrounding areas.

Based on the spatial distribution of slope angles, it has been seen that the 49.1% area possesses a slope between 0° and 0.4°, which is the most widespread. On the other hand, 23% area possesses a slope angle between 0.4° and 1.5°, whereas 18.5% is spread between 1.6° and 2.7°. Only 8% area possesses a slope angle between 2.8° and 4.8°. In general, the findings indicate that the study region is controlled by 90.6%, 0°–2.7°. The relationship between slope angle and shoreline erosion and inundation in the Indian Sundarbans is critical to understanding how susceptible the region is to environmental alterations (Hazra et al., 2002; Rudra & Chattopadhyay, 2019). The research areas account for very gradual slopes (0°–2.7°), are prone to increased shoreline erosion and flooding due to the low topographic features that render them more vulnerable to tidal surges, storms, and sea level rise (Idier et al., 2019; Xie et al., 2019). Wherever these points are found, coastal erosion accelerates, causing land to be lost and mangrove ecosystems that are important to protect against erosion to decline. Steeper slopes, on the other hand, exhibit high resistance against erosion as well as flooding but remain vulnerable during extreme weather conditions such as cyclones or tsunamis. The distinctive topography of the Indian Sundarbans, with its complex tidal flats and mangrove forests, is highly vulnerable to the interplay between slope angle, sea level rise, and increased storm frequency. This necessitates knowing and addressing these factors in order to carry out proper conservation and resilience planning.

1.1.3 Hydrometeorology Characteristic

The study area, located in a humid subtropical region, experiences a challenging summer from March to May. This is followed by a south-western monsoon that comes from June to September. The region is greatly affected by monsoon rains and tidal streams. It often faces cyclonic winds, floods, storms, and even submergence (Mondal et al., 2022a, b; Paul & Chaudhary, 2021). Seasonal changes, such as changes in temperature, separate rainfall, and annual air patterns (Mukherjee & Paul, 2021; Neogi et al., 2017), contribute significantly to the formation of tide and supra landscape. Bandyopadhyay (2019) indicates the Sundarbans' western side faces deadly south-western/southern winds in April/May, aligning with pre-monsoon cyclone seasons (like Fani, 2019) that inflict severe damage from these directions. The region experiences cyclones and high winds during this period, particularly impacting western coastal areas. On average, the region receives about 1663 mm of rainfall every year. The temperature varies from a cool 10 °C to a scorching 41 °C, with 75% (±5) (Biswas et al., 2017; Sahana et al., 2021) at the average level of humidity. Cyclonic storms and storm growth are common, especially during the pre-monsoon (April to May) and post-monsoon (October to November) seasons. These incidents pose serious risks for both local ecosystems and human communities (Mandal et al., 2024). The main watersheds are greatly influenced by the major rivers: Hugli, Matla, Bidyadhari, Raimangal, and Saptamukhi (Fig. 1.2) in this study area (Bhadra et al., 2017). Water is important to maintain this delicate ecosystem. It manages the flow of water through a complex web of rivers, estuaries, creeks, canals, and marshes (Anu et al., 2024; Liu et al., 2017). This process plays an important role in the influx of freshwater from upstream sources such as rivers and groundwater aquifers. It helps to recreate soil moisture levels, which are important for plant growth. In this delicate environment, sufficient moisture is necessary for plants to survive, especially in the high salinity of most soils of the GIV-En region (Mandal et al., 2019; Sudhir et al., 2022). Additionally, hydrological processes are necessary to maintain proper tidal flow, supporting a healthy fish population in these waterways (Saeed et al., 2025). Without effective water management, fish stocks can be severely affected by human activities, such as excessive pollution in local currents and lakes.

1.1.4 Soil, Groundwater, and Geological Characteristics

The soils in this region are predominantly sandy and contain low levels of organic matter, largely due to their proximity to the beach (Siddiqu et al., 2025; Das et al., 2024). These conditions pose significant challenges for coastal soils, particularly with respect to nitrogen availability and overall soil quality. Nitrogen levels may be reduced by 0.1%, which indicates poor soil condition (Sanyal et al., 2020; Beusen et al., 2022; Awoonor et al., 2025). These soils also contain high salinity and poor drainage. The flat landscape of the region often causes waterlogging during the monsoon season (Prajapati et al., 2021; Ghosh and Mistri, 2020). The average soil pH in the Indian Sundarbans is measured in all depths and years, 7.29. The western region indicates a high pH level (Biswas et al., 2017). The features of the soil can vary in different parts of the beauty, which are influenced by factors such as landform types, vegetative cover, and hydrogenotrophic conditions (Hossain et al., 2025a, b; Sahu et al., 2024). Sandy loam soils are prevalent, while silty clays can be found along riverbanks where deposition occurs due to river flow or tidal action from the nearby seas and oceans (Bucci et al., 2022; Park, 2024). Alluvial deposits, mainly from the Ganges and its tributaries, are common here (Bonsor et al., 2017; Frings et al., 2015).

The different soil types have been observed in this study area (Fig. 1.4) and play a crucial role in determining land suitability. Figure 1.4 shows that the Southwestern part of the Sundarbans, mainly the Sundarbans biosphere reserve, covers fine loamy soil. This area found that surface soils in the southern Sundarbans are predominantly medium-textured, mainly classified as loam and silt loam. The highest silt content was observed in areas influenced by stronger tidal currents from the Bay of Bengal, which affect sediment deposition near river-sea confluences (Siddique et al., 2025). The south-western part and the middle portions of the Indian Sundarbans cover fine alluvial soil. This region's soil is primarily fine alluvial, with some muddy swamps. It is slightly alkaline (pH 8.0–8.3) with salinity ranging from 4.5 to 5.8 ppt. Predominantly heavy-textured (silty clay to clay loam), the soil has low hydraulic conductivity, often leading to waterlogging that damages crops, especially in summer (Manna et al., 2024). They are richer in nutrients compared to Entisols, containing elements like calcium carbonate (lime) and higher levels of organic matter, thanks to periodic flooding over time (Siddique et al., 2025). The northern part of the Indian Sundarbans, especially Canning and Diamond Harbor, has been dominated by fine-fine loamy soil because these areas are part of the Gangetic Delta, where alluvial deposits from rivers have created fine-textured soils (Chatterjee et al., 2024). Overall, the study area, alluvium deposited by rivers, has formed deep, fine loamy to fine-textured soils, largely saline due to tidal seawater intrusion through creeks and sub-tributaries. These soils are imperfectly to poorly drained, with moderate to high salinity, remaining wet and saline for much of the year. They are best suited for salt-tolerant crops. The ecosystem is highly vulnerable to erosion, accretion, tidal surges, and severe natural disasters, which directly impact the topsoil of intertidal mudflats surrounding the islands. The average tidal amplitude is about 3.0 m, with some sea-facing islands experiencing amplitudes up to 5.0 m.

Aquifers are typically found within the Quaternary and Tertiary sediments deposited over time, with thicknesses varying from 75 m to as deep as 360 m, depending on location and sediment composition (Burman et al., 2019). The district's confined aquifers can be divided into two distinct groups. The upper group, located between 20 and 160 m deep, is characterized by a layer of sandy gravel at its base,

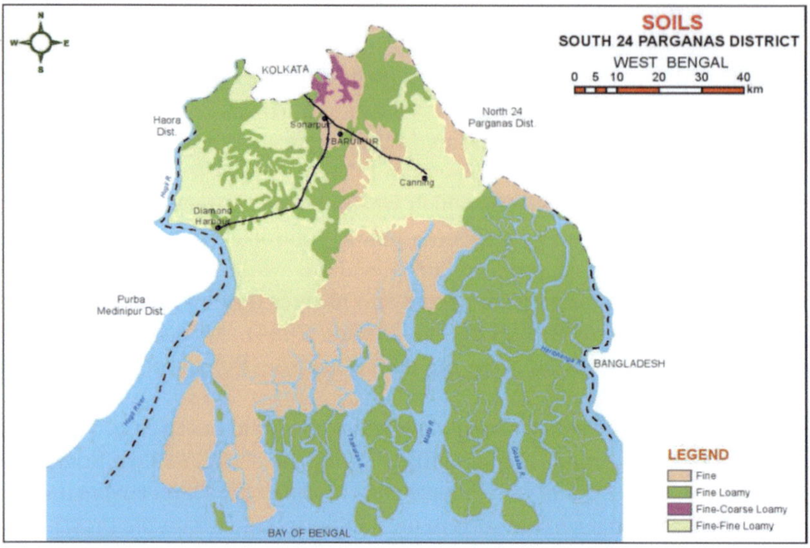

Fig. 1.4 Soil map of the Indian Sundarbans (*Source* NBSS & LUP Regional Centre, Calcutta)

which gradually fades toward the east. Groundwater in this layer is generally brackish to saline (chloride levels 1750–6300 ppm), making it unsuitable for drinking or irrigation. The lower group of aquifers, found at depths between 160 and 360 m, is a vital groundwater source. This productive aquifer is separated from the upper group by a thick, impermeable clay bed, which varies in thickness. According to the CGWB report from Kolkata (Gayen, 2009), the region's underground aquifers are divided into two main systems: Upper and Lower aquifers. These aquifers sit at average depths of about 130 m (±30) and 325 m (±25) below the surface. Several clay layers separate them, ranging from 40 to 150 m thick. In the area studied, water drawn from shallow to medium-depth tube wells (100–160 m bgl) is often brackish to saline (Gayen, 2009). This groundwater is mostly used for non-drinking purposes, like washing dishes. In contrast, the few deep tube wells (300–350 m bgl) provide relatively fresh water. Their electrical conductivity (EC) values range from 915 to 4,000 μmhos/cm at 25 °C, with chloride levels between 64 and 1,255 mg/l, making it suitable for drinking (Gayen, 2009). This difference in water quality creates a real challenge for local residents and farmers who need safe drinking water. Rainwater harvesting and conservation could be a practical solution, and this approach would let excess rainfall recharge shallow aquifers and gradually lower salinity.

The study area, located in the lower deltaic plain of India, is a fascinating region with unique features and situated on the composite Gangetic Delta. The area is covered by Quaternary sediments deposited (Fig. 1.5) by the Ganges River and its tributaries (Sarkar et al., 2009; Singh et al., 2014). The terrain is primarily composed of low-lying plains, with some areas featuring shallow depressions and marshy lands along the riversides. The sedimentary material, known as alluvium, is formed when water transports soil and rocks from higher elevations to lower ones. The top layer of this alluvium is primarily clay, with average thicknesses of 45 m (15–75 m), as reported by the Central Ground Water Board (2023).

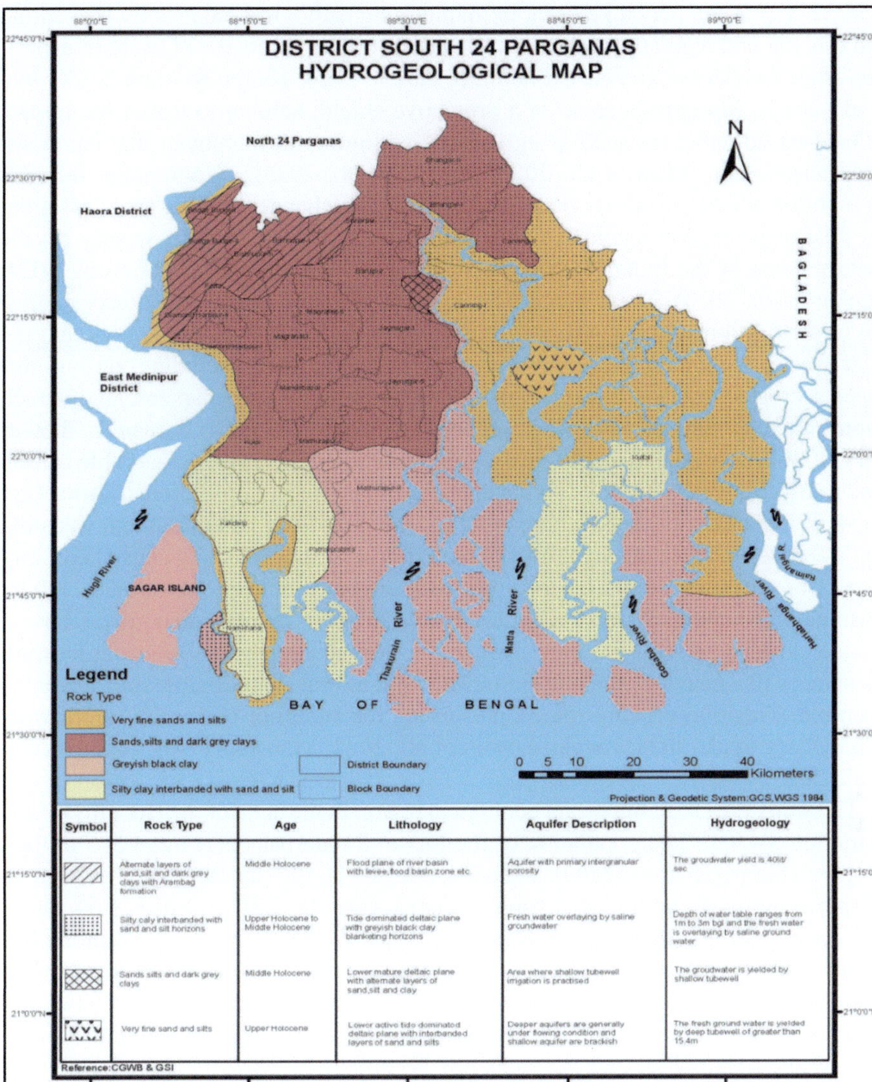

Fig. 1.5 Hydrology and Rock types of South 24 Parganas including study area (*Source* State Water Investigation Directorate, Govt. of West Bengal)

1.1.5 Mangroves and Demographic Conditions

The Sundarbans, the largest mangrove forests in India, are a treasure trove of biodiversity, teeming with a variety of plants and animals. These lush forests are not just beautiful; they also provide vital resources like timber, fuelwood, and honey that support the livelihoods of many people living nearby (Bansal et al., 2023; Ghosh et al., 2015). They play a key role in maintaining ecological balance by preventing soil erosion and regulating water availability through the process of evapotranspiration from their dense greenery (Saoum & Sarkar, 2024; Kanjin & Alam, 2024; Jain et al., 2023). Mangroves serve as a protective shield, helping to lessen the impact of land-based activities such as agriculture and urban development that can harm the environment (Khan et al., 2025; Sunkur et al., 2023). Additionally, they act as a buffer against rising sea levels due to climate change, making them essential for coastal protection against storms, surges, and cyclones that frequently hit the eastern coast of the Indian Subcontinent (Das & Swain, 2024; Dong et al., 2024; Sunkur et al., 2023). These mangroves are part of two interconnected ecosystems: the terrestrial forested areas next to estuarine wetlands, known as Mangal, and the tidal creeks within these regions that support marine life, including various fish species, crustaceans, and aquatic plants like seagrasses and salt marshes, all of which contribute to the unique Sundarbans ecosystem (Giri et al., 2007; Zaman & Chowdhury, 2024). Beyond their ecological significance, mangroves are remarkable carbon sinks, absorbing more carbon dioxide than any other type of tropical forest (Anu et al., 2024; Tengku Hashim & Suratman, 2021). This ability helps fight global warming and stabilizes local weather patterns, which is crucial for protecting coastal communities from extreme weather events related to climate change (Sunkur et al., 2023; Hasnine & Nagdeve, 2025; Rodrigues et al., 2024). Moreover, mangroves have a unique capacity to reduce human-induced pollution, as they can filter out harmful substances like heavy metals, oil spills, and agricultural runoff before they reach the open sea. This plays a vital role in preserving the health of our oceans worldwide (Choudhary et al., 2024; Awuku-Sowah et al., 2022; Moustafa et al., 2023; Akram et al., 2023).

The study area is an important ecological region in the delta of the Hooghly-Matla Estuarine System. The population of the Indian Sundarbans is about 4.5–5 million people. This population mainly includes Bengali Hindu and Muslims, as well as Christian and some indigenous tribal societies (Danda, 2007). Census data shows that the population of the Indian Sundarbans has significantly increased in the last 50 years (Rudra & Chattopadhyay, 2019). The region is spread to the south and north of the 24 Parganas districts in West Bengal and is mainly rural, with small cities and villages scattered. It is the range of many other Estuarine-Ease, including Muriganga, Saptamukhi, Thakuran, Gosaba, and Hasinbhanga. Low-lying on the western bank of an ancient delta formed by the Ganges River, AR-EA is mostly covered by mangrove forests that provide housing for various wildlife species. The economy mainly depends on agriculture, fishing, and aquaculture, although many people also collect honey and use resources from mangrove forests (Iqbal, 2020; Mallick et al.,

Table 1.1 Protected area (National Park) of South 24 Parganas, India

District	Flagship species	Area in km^2	Protected area (National Park)
S. 24 Parganas	Tiger	1330.10	Sundarbans N.P
	Tiger	362.40	Sajnakhali W.L.S
	Crocodile	5.9	Halliday Island W.L.S
	Crocodile	38.00	Lothian Island W.L.S

2021). Despite this, the region faces high levels of poverty, limited access to healthcare, and low education standards, which deteriorate with frequent natural disasters such as cyclones and floods (Ghosh & Mistri, 2023; Mondal et al., 2025). The Sundarbans Delta region also struggles with issues such as poor infrastructure, migration in urban centers for better opportunities, and the ongoing struggle needs forest protection and livelihood requirements (Jamal et al., 2022). Despite these challenges, the Sundarbans claim a rich cultural heritage, with lively local traditions, festivals, and close social lives. Table 1.1 shows the distribution of the protected forest area in South 24 Parganas, India.

1.1.6 Climate Change and Its Escalating Impact on Natural Hazards

Climate change is a global issue that has significantly increased sea temperature. Based on the report, Krishnan et al. (2020), during the year 1951–2015, the sea surface temperature (SST) of the tropical Indian Ocean rises from 1 °C, which crosses the global average SST growth of 0.7 °C. In the last six decades (1955–2015), the heat content of the ocean in the upper 700 m (OHC700) has shown a stable increase, with a sharp increase in the last two decades (1998–2015). In the Northern Indian Ocean (NIO), sea level growth took place from 1874 to 2004 at the rate of 1.06–1.75 mm per year. It has increased to 3.3 mm per year in the last two and a half decades from 1993 to 2017 (Krishnan et al., 2020). This rate is similar to the current global average marine-level growth. The heat material of SST and Ocean in the tropical Indian Ocean is expected to continue to grow during the twenty-first century (Fig. 1.6). This has affected the Indian Sundarbans region for several years (Datta et al., 2024; Ghosh & Mistri, 2023; Jabir et al., 2021). Climate change continues to influence the area, understanding the rise in frequency and intensity of cyclones is becoming more important due to rising sea levels (Carstens et al., 2025; Mulla et al., 2025; Venegas et al., 2023). The Bay of Bengal and the Arabian Sea are known hotspots for cyclone formation in the Indian Subcontinent, making them particularly susceptible to cyclones (Koley et al., 2024; Mondal et al., 2022a, b; Mulla et al., 2025; NCRMP, 2011; Venegas et al., 2023). It is reported that the North Indian Ocean (NIO) has been one of the most vulnerable regions prone to cyclonic storms, contributing to 7–10% of global cyclonic events during 1951 to 2021(Mondal et al.,

2021; Vellore et al., 2020; Koley et al., 2024). The Bay of Bengal is especially at risk, which severely affects coastal states like West Bengal, Odisha, Andhra Pradesh, and Tamil Nadu (Bandyopadhyay et al., 2021; Koley et al., 2024; Mondal et al., 2022a, b; Singh & Roxy, 2022).

The Sundarbans are more likely to experience tropical cyclones because of where they are located (Mojid et al., 2025). Their position along a warm ocean current in the Bay of Bengal, along with their closeness to low-pressure systems in the area, adds to this risk (Mishra et al., 2021; Wu et al., 2024). These conditions make it easier for cyclones to form and grow stronger, which is why the Sundarbans often face these intense storms. Since 1990, there has been a marked increase in both the intensity and frequency of cyclones affecting the area (Koley et al., 2024). Mangrove deforestation has further exacerbated coastal vulnerability, putting approximately 4.5 million people at risk from tropical cyclones in the Indian Sundarbans. Over the past two decades, the intensity and frequency of severe cyclonic storms in the region have surged (Koley et al., 2024), as shown in Table 1.2. In Sundarbans, a total of 40 cyclonic storms have been recorded along the West Bengal coast from 1924 to 2024, and 77 cyclonic storms have been observed along the Bangladesh Coast (Fig. 1.7a and b). Based on the frequency distribution of tropical cyclones (TCs) from 1951 to 2020, the highest frequency (25.34%) was observed in the Bangladesh Coastal region (Koley et al., 2024). In the Indian subcontinent, the Andhra Pradesh coast recorded the

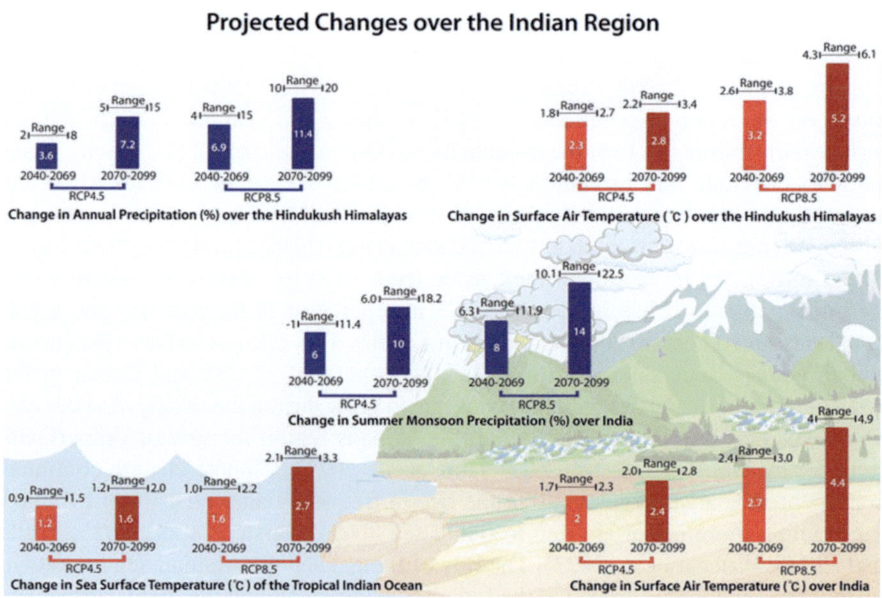

Fig. 1.6 Best estimate and range in climate model projections of future changes in 1. Surface air temperature over India 2. Sea surface temperature of the tropical Indian Ocean 3. Surface air temperature over the Hindu Kush Himalayas 4. Summer monsoon precipitation over India, 5. Annual precipitation over the Hindu Kush Himalayas (Krishnan et al., 2020)

Table 1.2 Major tropical cyclones (TCs) events in the Indian Sundarbans and its surroundings during 2009–2024

Cyclone name	Landfall date	Maximum wind speed (km/h)	Landfall locations	Storm surge height (m)	Damage (Million USD)
Aila	25/05/2009	120	Sagar Island, Indian Sundarbans	2–3	618
Mahasen	16/05/2013	85	NW of Chittagong, Bangladesh	2	107
Roanu	21/05/2016	110	NW of Chittagong, Bangladesh	3–5	2030
Bulbul	10/11/2019	155	Sagar Island and Frazerganj Indian Sundarbans	1–2	3370
Fani	03/05/2019	260	Puri, Orissa Coast	1.5	8100
Amphan	20/05/2020	270	Bakkhali, Indian Sundarbans	5.45	13,400
Yaas	23/05/2021	155	South of Balasore, Orissa	2–4	284
Remal	26/05/2024	120	Landfall between Bangladesh and West Bengal	3–4	90.7

highest frequency of TCs at 22.17%, followed by 15.84% on the Tamil Nadu coast, 14.93% on the Odisha coast, and 8.60% on the West Bengal coast along India's East Coast. It is observed that, comparatively, the Sundarbans part of Bangladesh is more vulnerable than the Indian part of the Sundarbans (Koley et al., 2024). In contrast, the number of very severe cyclonic storms (VSCSs) during the post-monsoon season has risen sharply, with an increase of more than one event per decade, over the last two decades (2000–2020) (Koley et al., 2024; Krishnan et al., 2020). These storms have devastated local communities, disrupting livelihoods and habitats, particularly mangrove forests, which serve as a natural buffer against storm surges, especially during the monsoon season when flooding is common (Ashrafuzzaman, 2025; Dong et al., 2024; Koley et al., 2024). In many cases, the damage to infrastructure and homes can take years to repair, if at all, leaving villages cut off from essential services and clean drinking water as their homes are swept away by the sea (Bolan et al., 2024; Dong et al., 2024; Tenhunen et al., 2023).

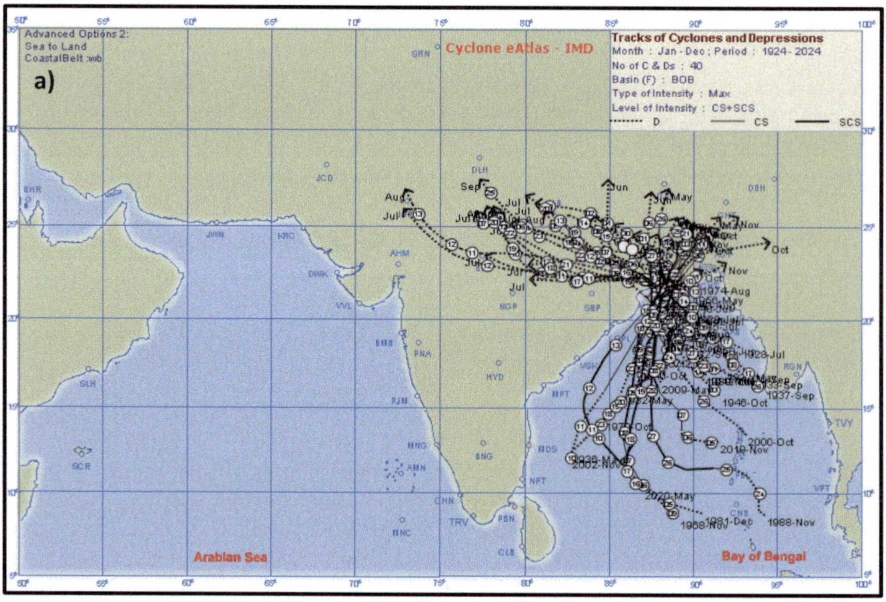

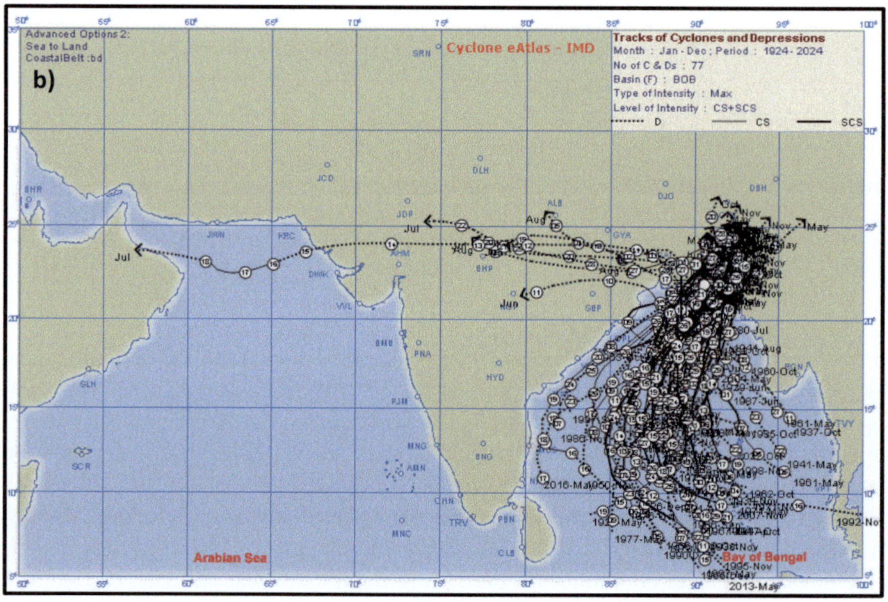

Fig. 1.7 Distribution of track of cyclonic storm in Sundarbans **a** West Bengal Coast, **b** Bangladesh Coast from 1924 to 2024 (*Source* Cyclone eAtlas-IMD)

References

Adnan, M.S.G., Dewan, A., Zannat, K. E. (2019). The use of watershed geomorphic data in flash flood susceptibility zoning: a case study of the Karnaphuli and Sangu river basins of Bangladesh. *Nat Hazards, 99*, 425–448. https://doi.org/10.1007/s11069-019-03749-3.

Akram, H., Hussain, S., Mazumdar, P., Chua, K. O., Butt, T. E., & Harikrishna, J. A. (2023). Mangrove health: A review of functions, threats, and challenges associated with mangrove management practices. *Forests, 14*(9), 1698. https://doi.org/10.3390/f14091698.

Ali, Y., & Rahman, M. M. (2025). Quantifying forest stocking changes in Sundarbans mangrove using remote sensing data. *Science of Remote Sensing, 11*, Article 100181. https://doi.org/10.1016/j.srs.2024.100181.

Anu, K., Sneha, V. K., & Busheera, P. (2024). Mangroves in environmental engineering: Harnessing the multifunctional potential of Nature's coastal architects for sustainable ecosystem management. *Results in Engineering, 21*, Article 101765. https://doi.org/10.1016/j.rineng.2024.101765.

Ashrafuzzaman, M. (2025). Mangroves as a cornerstone of nature-based solutions for global climate resilience. In: W. Leal Filho, G. J. Nagy, & D. Y. Ayal (Eds.), *Handbook of nature-based solutions to mitigation and adaptation to climate change.* Cham: Springer. https://doi.org/10.1007/978-3-030-98067-2_18-1.

Awoonor, J. K., Amoakwah, E., Buri, M. M., Dogbey, B. F., & Gyamfi, J. K. (2025). Impact of land use on soil quality: Insights from the forest-savannah transition zone of Ghana. *Heliyon, 11*(1). https://doi.org/10.1016/j.heliyon.2024.e41183.

Awuku-Sowah, E. M., Graham, N. A., & Watson, N. M. (2022). Investigating mangrove-human health relationships: A review of recently reported physiological benefits. *Dialogues in Health, 1*, Article 100059. https://doi.org/10.1016/j.dialog.2022.100059.

Bandyopadhyay, S., Dasgupta, S., Khan, Z. H., & Wheeler, D. (2021). Spatiotemporal analysis of tropical cyclone landfalls in Northern Bay of Bengal, India and Bangladesh. *Asia-Pacific Journal of Atmospheric Sciences, 57*, 799–815. https://doi.org/10.1007/s13143-021-00227-4.

Bandyopadhyay, S. (2019). Sundarban: A review of evolution and geomorphology. *World Bank Group, 36*. https://documents1.worldbank.org/curated/en/119121562735959426/pdf/Sundarban-A-Review-of-Evolution-and-Geomorphology.pdf.

Bansal, K., Choudhury, T., Nath, A., & Koley, B. (2023). NDVI indicator based land use/land cover change analysis using machine learning and geospatial techniques at Rupnarayan River Basin, West Bengal, India. In: V. S. Rathore, V. Piuri, R. Babo, M. C. Ferreira (Eds.), *Emerging trends in expert applications and security. ICETEAS 2023. Lecture notes in networks and systems* (Vol. 682). Singapore: Springer. https://doi.org/10.1007/978-981-99-1946-8_11.

Barbier, E., Hacker, S. D., Kennedy, C., Koch, E. W., Stier, A. C., & Silliman, B. R. (2011). The value of estuarine and coastal ecosystem services. *Ecological Monographs, 81*, 169–193.

Bera, R., & Maiti, R. (2019). Quantitative analysis of erosion and accretion (1975–2017) using DSAS—A study on Indian Sundarbans. *Regional Studies in Marine Science, 28*, Article 100583. https://doi.org/10.1016/j.rsma.2019.100583.

Bera, B., Bhattacharjee, S., & Sengupta, N. (2022). Significant reduction of carbon stocks and changes of ecosystem service valuation of Indian Sundarban. *Science and Reports, 12*, 7809. https://doi.org/10.1038/s41598-022-11716-5.

Beusen, A. H. W., Doelman, J. C., Van Beek, L. P. H., Van Puijenbroek, P. J. T. M., Mogollón, J. M., Van Grinsven, H. J. M., et al. (2022). Exploring river nitrogen and phosphorus loading and export to global coastal waters in the shared socio-economic pathways. *Global Environmental Change, 72*(2022), Article 102426. https://doi.org/10.1016/j.gloenvcha.2021.102426.

Bevington, A. E., Twilley, R. R., & Sasser, C. E. (2022). Deltaic floodplain wetland vegetation dynamics along the sediment surface elevation gradient and in response to disturbance from river flooding and hurricanes in Wax Lake Delta, Louisiana, USA. *Geomorphology, 398*, Article 108011. https://doi.org/10.1016/j.geomorph.2021.108011.

Bhadra, T., Mukhopadhyay, A., & Hazra, S. (2017). Identification of river discontinuity using geo-informatics to improve freshwater flow and ecosystem services in Indian Sundarban delta. In: S. Hazra, A. Mukhopadhyay, A. Ghosh, D. Mitra, V. Dadhwal (Eds.), *Environment and earth observation. Springer remote sensing/photogrammetry*. Cham: Springer. https://doi.org/10.1007/978-3-319-46010-9_10.

Bhuiyan, M. K. A., Godoy, O., González-Ortegón, E., Billah, M. M., & Rodil, I. F. (2025). Salt marsh macrofauna: An overview of functions and services. *Marine Environmental Research*, 106975. https://doi.org/10.1016/j.marenvres.2025.106975.

Biswas, S., Zaman, S., & Mitra, A. (2017). Soil characteristics of Indian Sunderbans: The designated world heritage site. *Scientific Journal of Biomedical Engineering & Biomedical Science, 1*(2), 053–059.

Bolan, S., Padhye, L. P., Jasemizad, T., Govarthanan, M., Karmegam, N., Wijesekara, H., et al. (2024). Impacts of climate change on the fate of contaminants through extreme weather events. *Science of the Total Environment, 909*, Article 168388. https://doi.org/10.1016/j.scitotenv.2023.168388.

Bonsor, H. C., MacDonald, A. M., Ahmed, K. M., et al. (2017). Hydrogeological typologies of the Indo-Gangetic basin alluvial aquifer, South Asia. *Hydrogeology Journal, 25*, 1377–1406. https://doi.org/10.1007/s10040-017-1550-z.

Bucci, M. G., Micallef, A., Urlaub, M., Mountjoy, J., & Barrett, R. (2022). A global review of subaqueous spreading and its morphological and sedimentological characteristics: A database for highlighting the current state of the art. *Geomorphology, 414*, Article 108397. https://doi.org/10.1016/j.geomorph.2022.108397.

Burman, D., Mondal, M. K., Khan, Z. H., Sutradhar, A. K., & Kamal, F. A. (2019). Soil and water resources of Sundarbans. In: H. Sen (Ed.), *The Sundarbans: A disaster-prone eco-region. Coastal research library* (Vol. 30). Cham: Springer. https://doi.org/10.1007/978-3-030-00680-8_6.

Carstens, J. D., Uejio, C. K., Powell, E., Jung, J., & Zonka, S. (2025). Tropical cyclones and climate change: An overview for the public health community. *Environmental Research*, 122149. https://doi.org/10.1016/j.envres.2025.122149.

CGWB. (2023). Ground water information booklet South 24 Parganas District, West Bengal. Government of India Ministry of Water Resources Central Ground Water Board. Retrieved February 12, 2023, from http://cgwb.gov.in/district_profile/westbangal/south_parganas.pdf.

Chatterjee, N., Koley, B., Nath, A., & Roy, S. (2024). Developing sustainable livelihood index for the coastal belt of Indian Sundarbans. In: T. Choudhury, B, Koley, A. Nath, J. S. Um, A. K. Patidar (Eds.), *Geo-environmental hazards using AI-enabled geospatial techniques and earth observation systems. Advances in geographic information science*. Cham: Springer. https://doi.org/10.1007/978-3-031-53763-9_13.

Choudhary, B., Dhar, V., & Pawase, A. S. (2024). Blue carbon and the role of mangroves in carbon sequestration: Its mechanisms, estimation, human impacts and conservation strategies for economic incentives. *Journal of Sea Research, 199*, Article 102504. https://doi.org/10.1016/j.seares.2024.102504.

Danda, A. A. (2007). *Surviving in the Sundarbans: Threats and responses* (p. 203). PhD Thesis—Research UT, graduation UT, University of Twente. https://ris.utwente.nl/ws/portalfiles/portal/14252552/surviving.pdf.

Das, A., & Swain, P. K. (2024). Navigating the sea level rise: Exploring the interplay of climate change, sea level rise, and coastal communities in India. *Environmental Monitoring and Assessment, 196*, 1010. https://doi.org/10.1007/s10661-024-13191-z.

Das, B.K., Chakraborty, H.J., Kumar, V. (2024). Comparative metagenomic analysis from Sundarbans ecosystems advances our understanding of microbial communities and their functional roles. *Scientific Report*, *14*, 16218. https://doi.org/10.1038/s41598-024-67240-1.

Datta, D., Chattopadhyay, R. N., & Deb, S. (2011). Prospective livelihood opportunities from the Mangroves of the Sunderbans, India. *Research Journal of Environmental Sciences, 5*, 536–543.

Datta, P., Behera, B., & Rahut, D. B. (2024). Climate change and water-related threats in the Indian Sundarbans: Food security and management implications. *International Journal of Water Resources Development, 40*(3), 323–344. https://doi.org/10.1080/07900627.2023.2224459.

Dong, W. S., Ismailluddin, A., Yun, L. S., Ariffin, E. H., Saengsupavanich, C., Maulud, K. N. A., et al. (2024). The impact of climate change on coastal erosion in Southeast Asia and the compelling need to establish robust adaptation strategies. *Heliyon, 10*(4). https://doi.org/10.1016/j.heliyon.2024.e25609.

Frings, P. J., Clymans, W., Fontorbe, G., Gray, W., Chakrapani, G. J., Conley, D. J., & De La Rocha, C. (2015). Silicate weathering in the Ganges alluvial plain. *Earth and Planetary Science Letters, 427*, 136–148. https://doi.org/10.1016/j.epsl.2015.06.049.

Gayen, A. (2009). Sustainable ground water management options in the delta mouth area of Sundarban Islands, West Bengal. *Bhujal Journal, 24*(1), 79–84.

Ghosh, A., Schmidt, S., Fickert, T., & Nüsser, M. (2015). The Indian Sundarban mangrove forests: History, utilization, conservation strategies and local perception. *Diversity, 7*(2), 149–169. https://doi.org/10.3390/d7020149.

Ghosh, S., Mistri, B. (2020) Drainage induced waterlogging problem and its impact on farming system: a study in Gosaba Island, Sundarban, India. Spatial Information Research, *28*, 709–721. https://doi.org/10.1007/s41324-020-00328-8.

Ghosh, S., & Mistri, B. (2023). Cyclone-induced coastal vulnerability, livelihood challenges and mitigation measures of Matla–Bidya inter-estuarine area, Indian Sundarban. *Natural Hazards , 116*, 3857–3878. https://doi.org/10.1007/s11069-023-05840-2.

Ghosh, S., & Roy, S. (2022). Climate change, ecological stress and livelihood choices in Indian Sundarban. In A. K. E. Haque, P. Mukhopadhyay, M. Nepal, M. R. Shammin (Eds.), *Climate change and community resilience* (pp. 399–413). Singapore: Springer. https://doi.org/10.1007/978-981-16-0680-9_26.

Giri, C., Pengra, B., Zhu, Z., Singh, A., & Tieszen, L. L. (2007). Monitoring mangrove forest dynamics of the Sundarbans in Bangladesh and India using multi-temporal satellite data from 1973 to 2000. *Estuarine, Coastal and Shelf Science, 73*(1–2), 91–100. https://doi.org/10.1016/j.ecss.2006.12.019.

Hasnine, M., & Nagdeve, D. A. (2025). Natural hazard induced coastal vulnerability in Indian Sundarbans: A village-level study by using geospatial and statistical techniques. *Natural Hazards Research, 5*(2), 262–275. https://doi.org/10.1016/j.nhres.2024.10.004.

Hazra, S., Ghosh, T., DasGupta, R., & Sen, G. (2002). Sea level and associated changes in the Sundarbans. *Science and Culture, 68*(9/12), 309–321.

Hossain, M. T., Shohel, T. A., Ahsan, M. N., & Islam, M. N. (2025b). Prevalence and determinants of vulnerability among Sundarbans mangrove forest resource-dependent communities in cyclone-prone southwestern coastal districts of Bangladesh. *Trees, Forests and People, 19*, Article 100739. https://doi.org/10.1016/j.tfp.2024.100739.

Hossain, M. K., Kabir, M. A., Molla, M. O. F., Hossain, M. A., Khatun, M. S., Lima, M. N. J., & Islam, R. (2025). Geospatial analysis of soil salinity dynamics: Exploring topographic and vegetation influences in coastal Bangladesh. *Geosystems and Geoenvironment*, 100418. https://doi.org/10.1016/j.geogeo.2025.100418.

Idier, D., Bertin, X., & Thompson, P. (2019). Interactions between mean sea level, tide, surge, waves and flooding: Mechanisms and contributions to sea level variations at the coast. *Surveys in Geophysics, 40*, 1603–1630. https://doi.org/10.1007/s10712-019-09549-5.

Iqbal, M. H. (2020). Valuing ecosystem services of Sundarbans mangrove forest: Approach of choice experiment. *Global Ecology and Conservation, 24*, Article e01273. https://doi.org/10.1016/j.gecco.2020.e01273.

Islam, S. N. (2016). Deltaic floodplains development and wetland ecosystems management in the Ganges–Brahmaputra–Meghna Rivers delta in Bangladesh. *Sustain. Water Resour. Manag., 2*, 237–256. https://doi.org/10.1007/s40899-016-0047-6.

Jabir, A. A., Hasan, G. J., & Anam, M. M. (2021). Correlation between temperature, sea level rise and land loss: An assessment along the Sundarbans coast. *Journal of King Saud University-Engineering Sciences*. https://doi.org/10.1016/j.jksues.2021.07.012.

Jain, A., Nath, A., Koley, B., Choudhury, T., & Um, J. S. (2023). Object-based vegetation mapping in the Sundarbans using machine learning techniques and earth observation data. In A. Ramdane-Cherif, T. P., Singh, R. Tomar, T. Choudhury, J. S. Um (Eds.), *Machine intelligence and data science applications. MIDAS 2022. Algorithms for intelligent systems*. Singapore: Springer. https://doi.org/10.1007/978-981-99-1620-7_42.

Jamal, S., Ghosh, A., Hazarika, R., & Sen, A. (2022). Livelihood, conflict and tourism: An assessment of livelihood impact in Sundarbans, West Bengal. *International Journal of Geoheritage and Parks, 10*(3), 383–399. https://doi.org/10.1016/j.ijgeop.2022.07.004.

Kanjin, K., & Alam, B. M. (2024). Assessing changes in land cover, NDVI, and LST in the Sundarbans mangrove forest in Bangladesh and India: A GIS and remote sensing approach. *Remote Sensing Applications: Society and Environment, 36*, Article 101289. https://doi.org/10.1016/j.rsase.2024.101289.

Kar, N. S., & Basu, A. (2023). A cyclone preparedness plan for the coastal blocks of the Indian Sundarban. *Journal of Coastal Conservation, 27*, 49. https://doi.org/10.1007/s11852-023-00978-4.

Khan, M. S., Akter, A., Khan, I., Aguila, L. C. R., Akter, L., Yang, M., et al. (2025). Assessing long term impact of regional climate and human activities on vegetation greenness dynamics in Sundarbans. *Ecological Informatics, 85*, Article 102984. https://doi.org/10.1016/j.ecoinf.2024.102984.

Khojasteh, D., Glamore, W., Heimhuber, V., & Felder, S. (2021). Sea level rise impacts on estuarine dynamics: A review. *Science of the Total Environment, 780*, Article 146470. https://doi.org/10.1016/j.scitotenv.2021.146470.

Koley, B., Nath, A., & Choudhury, T. (2024). Spatiotemporal trends of tropical cyclones in Bay of Bengal Basin, India. In T. Choudhury, B. Koley, A. Nath, J. S. Um, A. K. Patidar (Eds.), *Geo-environmental hazards using AI-enabled geospatial techniques and earth observation systems. Advances in geographic information science*. Cham: Springer. https://doi.org/10.1007/978-3-031-53763-9_3.

Krishnan, R., Sanjay, J., Gnanaseelan, C., Mujumdar, M., Kulkarni, A., & Chakraborty, S. (2020). Assessment of climate change over the Indian region: A report of the ministry of earth sciences (MOES). In *Government of India* (p. 226). Springer Nature. https://doi.org/10.1007/978-981-15-4327-2.

Liu, Q., Mou, X., Cui, B., & Ping, F. (2017). Regulation of drainage canals on the groundwater level in a typical coastal wetlands. *Journal of Hydrology, 555*, 463–478. https://doi.org/10.1016/j.jhydrol.2017.10.035.

Mahanta, B. N., Kashyap, M. P., & Mahapatra, B. M. (2024). Tectonic anabranching of Brahmaputra River system: Implications on survival of world's largest inhabited river island Majuli. *Discover Geoscience, 2*, 93. https://doi.org/10.1007/s44288-024-00100-7.

Mallick, B., Priodarshini, R., Kimengsi, J. N., Biswas, B., Hausmann, A. E., Islam, S., Huq, S., & Vogt. J. (2021). Livelihoods dependence on mangrove ecosystems: Empirical evidence from the Sundarbans. *Current Research in Environmental, Sustainability, 3*: 100077. https://doi.org/10.1016/j.crsust.2021.100077.

Mahmood, H., Ahmed, M., Islam, T., Uddin, M. Z., Ahmed, Z. U., & Saha, C. (2021). Paradigm shift in the management of the Sundarbans mangrove forest of Bangladesh: Issues and challenges. *Trees, Forests and People, 5*, Article 100094. https://doi.org/10.1016/j.tfp.2021.100094.

Mandal, U. K., Burman, D., Bhardwaj, A. K., Nayak, D. B., Samui, A., Mullick, S., et al. (2019). Waterlogging and coastal salinity management through land shaping and cropping intensification in climatically vulnerable Indian Sunderbans. *Agricultural Water Management, 216*, 12–26. https://doi.org/10.1016/j.agwat.2019.01.012.

Mandal, P., Roy, P., Maiti, A., Yadav, A., Bhattacharya, S., & Paul, S. (2024). Assessing the interplay of environmental change, cyclonic hazards, and their ramifications on agriculture in Indian

Sundarban. *Ocean & Coastal Management, 251*, Article 107080. https://doi.org/10.1016/j.ocecoaman.2024.107080.

Manna, M., Singh, D. K., Hasan, M., et al. (2024). Spatial variation of soil and ground water salinity in a deltaic region of the Indian Sundarbans. *Journal of Coastal Conservation, 28*, 41. https://doi.org/10.1007/s11852-024-01043-4.

Mishra, M., Acharyya, T., Santos, C. A. G., da Silva, R. M., Kar, D., Kamal, A. H. M., & Raulo, S. (2021). Geo-ecological impact assessment of severe cyclonic storm Amphan on Sundarban mangrove forest using geospatial technology. *Estuarine, Coastal and Shelf Science, 260*, Article 107486.

Mojid, M. A., Mainuddin, M., Karim, F., & Wahid, S. M. (2025). Historical and projected future hydrological characteristics of the mangrove forest in the Ganges Delta—A review. *Water, 17*(6), 838. https://doi.org/10.3390/w17060838.

Mondal, M., Haldar, S., Biswas, A., Mandal, S., Bhattacharya, S., & Paul, S. (2021). Modeling cyclone-induced multi-hazard risk assessment using analytical hierarchical processing and GIS for coastal West Bengal. *India Reg Stud Mar Sci, 44*, Article 101779. https://doi.org/10.1016/j.rsma.2021.101779.

Mondal, M., Biswas, A., Haldar, S., Mandal, S., Bhattacharya, S., & Paul, S. (2022a). Spatio-temporal behaviours of tropical cyclones over the Bay of Bengal Basin in last five decades. *Tropical Cyclone Research and Review, 11*(1), 1–15. https://doi.org/10.1016/j.tcrr.2021.11.004.

Mondal, M., Biswas, A., Haldar, S., Mandal, S., Mandal, P., Bhattacharya, S., & Paul, S. (2022b). Climate change, multi-hazards and society: An empirical study on the coastal community of Indian Sundarban. *Natural Hazards Research, 2*(2), 84–96. https://doi.org/10.1016/j.nhres.2022.04.002.

Mondal, I., Ariful Hossain, S., Bandyopadhyay, J. (2025). Potential effects of projected global sea level rise on Sundarbans mangrove wetland ecosystem: insights from SLAMM and hybrid machine learning models. *Environment Development Sustainability, 27*, 14879–14912. https://doi.org/10.1007/s10668-025-06193-5.

Moustafa, A. A., Abdelfath, A., & Arnous, M. O. (2023). Monitoring temporal changes in coastal mangroves to understand the impacts of climate change: Red Sea, Egypt. *Journal of Coastal Conservation, 27*, 37. https://doi.org/10.1007/s11852-023-00970-y.

Mukherjee, D., & Paul, A. K. (2021). Mangrove sensitivities to climate change and its impacts in the Sundarbans: A case study in the Patibania Island of South Western Sundarbans, India. In *Modern cartography series* (Vol. 10, pp. 353–385). Academic Press. https://doi.org/10.1016/B978-0-12-823895-0.00030-0.

Mulla, S., Singh, S. K., & Ahmed, R. (2025). Assessing the impact of climate change on land-falling tropical cyclones (LFTCs) over the North Indian Ocean (NIO) and their effects on coastal agriculture in Maharashtra: A case study. *Tropical Cyclone Research and Review*. https://doi.org/10.1016/j.tcrr.2025.04.003.

NCRMP. (2011). *National cyclone risk mitigation project*. The Hindu. Retrieved 18 February, 2023, from http://www.thehindu.com/2011/01/29/stories/2011012966692000.html.

Neogi, S. B., Dey, M., Kabir, S. L., Masum, S. J. H., Kopprio, G., Yamasaki, S., & Lara, R. (2017). Sundarban mangroves: Diversity, ecosystem services and climate change impacts. *Asian Journal of Medical and Biological Research, 2*(4), 488–507. https://doi.org/10.3329/ajmbr.v2i4.30988.

Park, E. (2024). Sand mining in the Mekong Delta: Extent and compounded impacts. *Science of the Total Environment, 924*, Article 171620. https://doi.org/10.1016/j.scitotenv.2024.171620.

Paul, S., & Chowdhury, S. (2021) Investigation of the character and impact of tropical cyclone Yaas: A study over coastal districts of West Bengal, India. *Safety in Extreme Environments, 3*, 219–235. https://doi.org/10.1007/s42797-021-00044-y.

Prajapati, G. S., Rai, P. K., Mishra, V. N., Singh, P., & Shahi, A. P. (2021). Remote sensing-based assessment of waterlogging and soil salinity: A case study from Kerala, India. *Results in Geophysical Sciences, 7*, Article 100024. https://doi.org/10.1016/j.ringps.2021.100024.

Rashid, M. B., Habib, M. A., Sultan-Ul-Islam, M., Khan, R., & Islam, A. R. M. T. (2024). Synthesis of drainage characteristics, water resources and sediment supply of the Bengal basin. *Quaternary Science Advances, 16*, 100244. https://doi.org/10.1016/j.qsa.2024.100244.

Rodrigues, J. N., Bhattacharya, S., Sachdev, B. K., & Cabete, D. (2024). Disaster management in the Sundarbans delta ecosystem through environmental governance and sustainable planning. *Revista De Direito Econômico e Socioambiental, 15*(2), 9. https://doi.org/10.7213/revdireconsoc.v15i2.31364.

Rudra, A., & Chattopadhyay, A. (2019). Environmental change of coastal Sundarbans: Impact on livelihood and standard of living status of indigenous people. *Environmental Quality Management, 29*(2), 77–84. https://doi.org/10.1002/tqem.21667.

Saeed, E., Al-Amir, N. & Elfeki, A. (2025). Assessment of dams' hydrological efficiency under the effect of climate change and urban expansion: Brayman Dam in Jeddah City, Saudi Arabia as a case study. *Natural Hazards, 121*, 6233–6255. https://doi.org/10.1007/s11069-024-07032-y.

Sahana, M., Ahmed, R., & Sajjad, H. (2016). Analyzing land surface temperature distribution in response to land use/land cover change using split window algorithm and spectral radiance model in Sundarban Biosphere Reserve, India. *Modeling Earth Systems and Environment, 2*, 81. https://doi.org/10.1007/s40808-016-0135-5.

Sahana, M., Rehman, S., & Ahmed, R. (2021). Analyzing climate variability and its effects in Sundarban Biosphere Reserve, India: Reaffirmation from local communities. *Environment, Development & Sustainability, 23*, 2465–2492. https://doi.org/10.1007/s10668-020-00682-5.

Sahu, H., Purohit, P., Srivastava, A., Singh, R., Mishra, A. P., Arunachalam, K., & Kumar, U. (2024). Comparative assessment of soil parameters and ecological dynamics in the Western Himalayan wetland and its surrounding periphery. *Environmental Quality Management, 34*(1), Article e22283. https://doi.org/10.1002/tqem.22283.

Sanyal, P., Ray, R., Paul, M., Gupta, V. K., Acharya, A., Bakshi, S., et al. (2020). Assessing the dynamics of dissolved organic matter (DOM) in the coastal environments dominated by mangroves, Indian Sundarbans. *Frontiers in Earth Science, 8*, 218. https://doi.org/10.3389/feart.2020.00218.

Saoum, M. R., & Sarkar, S. K. (2024). Monitoring mangrove forest change and its impacts on the environment. *Ecological Indicators, 159*, Article 111666. https://doi.org/10.1016/j.ecolind.2024.111666.

Sardar, P., & Samadder, S. R. (2021). Understanding the dynamics of landscape of greater Sundarban area using multi-layer perceptron Markov chain and landscape statistics approach. *Ecological Indicators, 121*, Article 106914. https://doi.org/10.1016/j.ecolind.2020.106914.

Sarkar, A., Sengupta, S., McArthur, J. M., Ravenscroft, P., Bera, M. K., Bhushan, R., et al. (2009). Evolution of Ganges-Brahmaputra western delta plain: Clues from sedimentology and carbon isotopes. *Quaternary Science Reviews, 28*(25–26), 2564–2581. https://doi.org/10.1016/j.quascirev.2009.05.016.

Sarkar, S. K. (2022). The imperiled Sundarban: Wetland of international importance. In D. A. DellaSala, & M. I. Goldstein (Eds.), *Imperiled: The Encyclopedia of conservation* (pp. 881–886). Elsevier. https://doi.org/10.1016/B978-0-12-821139-7.00092-1.

Siddique, A. B., Islam, M., Tasnim, N., Al Maruf, A., Al Helal, A. S., & Howlader, M. (2025). Spatial distribution of soil organic carbon and nutrients in the southern Sundarbans of Bangladesh. *Watershed Ecology and the Environment* (Vol. 7, p. 310–326). https://doi.org/10.1016/j.wsee.2025.03.005.

Singh, V. K., & Roxy, M. K. (2022). A review of ocean-atmosphere interactions during tropical cyclones in the north Indian Ocean. *Earth Science Reviews, 226*, Article 103967. https://doi.org/10.1016/j.earscirev.2022.103967.

Singh, N., Singh, R. P., Mukherjee, S., McDonald, K., & Reddy, K. J. (2014). Hydrogeological processes controlling the release of arsenic in parts of 24 Parganas district, West Bengal. *Environmental Earth Sciences, 72*(1), 111–118.

Sreelekshmi, S., Nandan, S. B., Kaimal, S. V., Radhakrishnan, C. K., & Suresh, V. R. (2020). Mangrove species diversity, stand structure and zonation pattern in relation to environmental factors—A case study at Sundarban delta, east coast of India. *Regional Studies in Marine Science, 35*, 101111. https://doi.org/10.1016/j.rsma.2020.101111.

Sudhir, S., Arunprasath, A., & Vel, V. S. (2022). A critical review on adaptations, and biological activities of the mangroves. *Journal of Natural Pesticide Research, 1*, Article 100006. https://doi.org/10.1016/j.napere.2022.100006.

Sunkur, R., Kantamaneni, K., Bokhoree, C., & Ravan, S. (2023). Mangroves' role in supporting ecosystem-based techniques to reduce disaster risk and adapt to climate change: A review. *Journal of Sea Research, 196*, Article 102449. https://doi.org/10.1016/j.seares.2023.102449.

Tengku Hashim, T. M. Z., & Suratman, M. N. (2021). Mangroves as a carbon sink/stocks. In: R. P. Rastogi, M. Phulwaria, D. K. Gupta (Eds.), *Mangroves: Ecology, biodiversity and management*. Singapore: Springer. https://doi.org/10.1007/978-981-16-2494-0_7.

Tenhunen, S., Uddin, M. J., & Roy, D. (2023). After cyclone Aila: Politics of climate change in Sundarbans. *Contemporary South Asia, 31*(2), 222–235. https://doi.org/10.1080/09584935.2023.2203903.

UNESCO (2021). Technical Guidelines for Biosphere Reserves, a new tool for the MAB Programme. Available online at: https://en.unesco.org/news/technical-guidelines-biosphere-reserves-new-tool-mab-programme. Accessed 05 Feb 2024.

Vellore, R. K., Deshpande, N., Priya, P., Singh, B. B., Bisht, J., Ghosh, S. (2020). Extreme storms. In: R. Krishnan, J. Sanjay, C. Gnanaseelan, M. Mujumdar, A. Kulkarni, & S. Chakraborty (Eds.), *Assessment of climate change over the Indian Region*. Singapore: Springer. https://doi.org/10.1007/978-981-15-4327-2_8.

Venegas, R. M., Acevedo, J., & Treml, E. A. (2023). Three decades of ocean warming impacts on marine ecosystems: A review and perspective. *Deep Sea Research Part II: Topical Studies in Oceanography, 212*, Article 105318. https://doi.org/10.1016/j.dsr2.2023.105318.

Warren-Rhodes, K., Schwarz, A. M., Boyle, L. N. G., Albert, J., Agalo, S. S., Warren, R., et al. (2011). Mangrove ecosystem services and the potential for carbon revenue programmes in Solomon Islands'. *Environmental Conservation, 38*(4), 485–496. https://doi.org/10.1017/S0376892911000373.

WBMDTCL. (2018). District survey report of minor mineral of South 24 Parganas district. Prepared as per Ministry of Environment, Forest and Climate Change Notification S. O. 3611 (E), Dated 25th July 2018. https://s24pgs.gov.in/assets/webdoc/mediaupload/x24d2d7b4a520fc497650fc11971cfa5e.pdf.

Wu, D., Ju, X., Sun, J., Hu, X., & Xiong, X. (2024). Escalation of tropical cyclone impacts on the Northwestern Bay of Bengal over the past decade. *Climate Dynamics, 62*(6), 5645–5662. https://doi.org/10.1007/s00382-024-07252-9.

Xie, D., Zou, Q. P., Mignone, A., & MacRae, J. D. (2019). Coastal flooding from wave overtopping and sea level rise adaptation in the Northeastern USA. *Coastal Engineering, 150*, 39–58. https://doi.org/10.1016/j.coastaleng.2019.02.001.

Zaman, M. S., & Chowdhury, T. H. (2024). The Sundarbans, the world's largest tidal halophytic mangrove forest: Its economic and ecological significance. *Bangla Journal of Interdisciplinary Sciences, 2*(1), E1–E15.

Chapter 2
Geo-Informatics Applications on Shoreline Shifting Analysis

Abstract The focus of this chapter is on the application of various GIS in coastal research and study based on a literature review. Scientists have emphasized the use of GIS for prediction and impact on the marine environment. Spatial information has been obtained for the sustainable management of the coastal region encompassed by the sea. In the current condition of data acquisition, management, and analysis are the main required factors for rapid accuracy and economical solutions. A GIS application is helpful for assessing the future condition of the shoreline based on the past and present scenarios of any given region. Digital Shoreline Analysis System is an extension tool of ArcGIS and is formed by a statistical algorithm model. This chapter emphasizes the wide application of remote sensing, GIS, and DSAS tools for assessing the shoreline change rate as well as the zonation map construction by obtaining values from shoreline analysis. This chapter has explored the versatile use of GIS and the DSAS operating system for coastal research.

Keywords Shoreline Shifting Analysis · Remote Sensing and GIS · Digital Shoreline Analysis System (DSAS) · Geospatial Techniques · Indian Sundarban

2.1 Introduction

In the past 30–40 years, Geographical Information System (GIS) was only applied to show the various land use research work, but in the last 15–20 years, GIS has been widely used in coastal or marine sciences (Goodchild, 2000). Support for the study of the global industrial system and the impact of anthropogenic environmental change has grown steadily. Geographers have begun to develop their focus on research boundaries, with traditional boundaries being defined in the 1970 and 1980s. Coastal GIS gained significant importance in the 1990s with the advent of systems science and an interdisciplinary approach to understand the entire Earth System globally, and with the growth of the population (Nierenberg, 1992; Williamson, 1994).

Other agents that increased the popularity of coastal geography and eventually its inclusion in GIS incorporate global environmental understanding and an increased

© The Author(s), under exclusive license to Springer Nature Switzerland AG 2026
A. Nath et al., *Shoreline Modification Over the Indian Sundarban*,
SpringerBriefs in GIS, https://doi.org/10.1007/978-3-032-13981-8_2

awareness of the significance of coastal ecology in worldwide conservation. From the United Nations convention, the new technologies for coastal research have been adopted (Psuty et al., 2002). Defining the dynamics of coastal ecosystems is essential for interpreting the correlation between coastal environments and species at the core of any initiative for marine area management. The development of spatial ideas across several aspects of the research methodology applied GIS in the coastal environment becomes very crucial, from data collection to decision-making and development by the GIS applications. Various researchers have applied spatial cognition in the past (Blades, 1991; Golledge & Stimson, 1987; Lloyd, 1997; Nyerges, 1991; Slater, 1982). Another key to establishing the spatial ideas is understanding various data models as well as applications in the GIS platform (Blades, 1991; Cances et al., 2000; Lloyd, 1997).

A GIS application is a versatile operating platform for coastal research. In GIS applications, data has been acquired fulfilling the aim of defining the modified nature of the coastal environment with various principles and methods, with several technical applications, though GIS can collect multiple data related to the earth surface system and also understand the impact of the coastal environment (Valavanis, 2002), as well as the spatio-temporal characteristics of the obtained data.

GIS is developed by the combination of an integrated management system with analytical processes of technology. Coastal GIS has been applied in various ways, i.e., data acquiring, mapping and monitoring tool with several discipline like coastal zone management, assessing the zone, land formation process, coastal geomorphology, coastal biodiversity, coastal population, climate change, sea level rise, flooding, erosion-accretion zonation mapping, saltwater intrusion, eutrophication in coastal zone, oil spill and pollution and policy-making and natural hazard disaster assessment with environmental sensitivity mapping (Charlier, 2000; Valavanis, 2002). It is crucial to understand the negative impact on the coast and also require the proper intervention. To develop the proper sustainable practices and fruitful policies in coastal zones, proper monitoring and various modeling can be helpful. With the help of GIS, this monitoring and modeling of various effects on the coastal zone can be assessed very quickly. GIS can provide better assistance for decision-making policies to increase awareness among the native people and sustainable eco-tourism in the coastal zone (ESRI, 2014), as well as maintain the water quality.

2.2 Literature Review of GIS on Coastal Research

Researchers in the past have stressed the significance of coastal zone assessment through GIS, defining the difference between marine applications and terrestrial GIS to protect the coastal areas from several triggering factors (Hamre, 1993; Li & Saxena, 1993), in coastal management. GIS data helps to evaluate the spatial characteristics (Lucas et al., 1994). A GIS-based model was developed by Beusen et al. (1995) to estimate the nitrogen and phosphorus load in European coastal areas through integrated hydrological data. In the past, several researchers have introduced the measurement method of uncertainty and the sensitivity analysis process through GIS (Crosetto et al., 2000; Gahegan & Ehlers, 2000; Hwang et al., 1998).

A management and monitoring-based GIS model was developed by Li et al. (1998) for Malaysia's coastline. This tool has been performing with the combination of time series data and spatial data. An international research program on earth sciences was conducted, where the GIS and Remote sensing data modeling was applied for in situ measurement and coastal environment monitoring (Van Zuidam et al., 1998). Li et al. (2001) used the GIS and remote sensing analysis method for assessing the shoreline change with high-resolution satellite imagery along the Lake Erie South shoreline.

To prepare the sensitivity mapping or index in GIS, Garcia et al. (2000) incorporated the anthropogenic factors into the earth surface phenomenon along Southern Spain. A physical sensitivity zonation mapping in the UK coastline and assessment of human health was analyzed by MacDonald and Cain (2000) due to the shipping in the region.

2.3 Application of GIS and Remote Sensing in Various Coastal Research

Currently, the coastal region has been experiencing intense physical and human-induced disturbances, including coastal flooding, sea level rise, coastal erosion, shoreline shifting, storm surges, and other various hazards, as well as mangrove loss. In the recent era of coastal research, shoreline shifting and erosion-accretion mapping have been widely prepared based on GIS and remote sensing applications with the shoreline change rate (Kumar & Jayappa, 2009; Ryu et al., 2002; Siddiqui & Maajid, 2004; Singh, 1989; White & El Asmar, 1999; Yamano et al., 2006).

The present research work stressed the shoreline modification condition at the Indian Sundarbans using GIS and remote sensing, and found the condition of land use and land cover. GIS and remote sensing are the most reliable and cost-effective applications with the versatility of analytical tools for coastal research. With the reliable accuracy of coastal changes and monitoring, remote sensing is widely applied with the integrated GIS method (Emran et al., 2016; Wang et al., 2013; Yanli, 2011). In several neo-tectonic studies, remote sensing data and GIS have also been applied and used in catchment area demarcation or to identify the basin in river areas (Islam & Hasan, 2020; Maurya et al., 2010, 2021).

GIS applications provides the largest number of accuracy through multi-resolution satellite data in a spatio-temporal condition to assess the shoreline modification rate applied by several recent researchers (Cracknell et al., 1982; Wagner et al., 1991; Nayak & Sahai, 1985; Ahmed & Neil, 1994; Ramalingam & Renganathan, 1998; Nayak, 2002; Addo & Kufogbe, 2011; Kabuth et al., 2013; Anand et al., 2016; Nandi et al., 2016; Masria et al., 2015; Hill et al., 1994; Murali et al., 2015; Tran Thi et al., 2014; Nassar et al., 2018; Suresh & Sundar, 2011; Li & Damen, 2010; Cui & Li, 2011; Lian & Jianfei, 2011; Durduran, 2010).

To acquire the periodical shifting of shoreline, multi-resolution satellite data has been applied, reaching high accuracy by Mujabar and Chandrasekar (2011). Another study reveals that satellite image provides the highest accuracy and accurate coastline position for modification detection (Saranathan et al., 2011). A study defines that remote sensing is the most effective application method to evaluate the shoreline position and estimate the regression and transgression rate (Marghany et al., 2010).

2.4 Digital Shoreline Analysis System (DSAS)

The DSAS application is an important tool for USGS coastal change risk projection, as it has standard and dynamic as well as simple iterative methods for assessing and studying variable morphology. It is an extension of ArcGIS software that was developed by the United States Geological Survey (USGS) (Himmelstoss et al., 2009). This software provides pre-examined statistical functions that allow for computing the multi-resolution shorelines as a time series of data with a chosen interval by the user for assessing the shoreline modification rate of any coastal region. This tool of DSAS provides a very reliable transect method for the shoreline change analysis (Figs. 2.1 and 2.2). In the DSAS algorithm, a baseline has been required for analysis. Generally, a baseline has been constructed by the user, and it is considered the starting point of casting each transect (Leatherman & Clow, 1983). Transects are intersected with the shoreline, and the intersection point is considered for the change rate calculation as an algorithm in the ArcGIS software. Baseline can be generated anywhere, i.e., onshore, offshore, or mid-shore position of the shoreline. To generate the baseline for calculation, some adequate fields have been required, i.e., OBJECTID, SHAPE, SHAPE_Length, ID, Group, and Search_Distance. Primarily, a single baseline has been drawn from the coastline before generating the transect on the statistics through either a specific distance or a buffer method from the beach (Nandi et al., 2016). The DSAS performs through five statistical algorithms for assessing the shoreline change rate, i.e., Shoreline Change Envelope (SCE), End Point Rate (EPR), Net Shoreline Movement (NSM), Linear Regression Rate (LRR), and Least Median of Squares (LMS) on long-term and short-term scales. Each algorithm has been calculating the change rate based on differences between two successive shoreline positions with a time interval. Obtained rates are reaching out as the unit of meters along the transects.

The model of NSM has been calculated based on the distance of each transect of the oldest and youngest shorelines (Fig. 2.3). On the other hand, SCE considers the distance between shorelines but not as a rate. SCE reports the highest distance of a transect along the intersection point. On the other hand, the EPR calculates two shoreline distances by dividing each other and obtains the rate in meters. Most current and the oldest shorelines have been divided from each other in the case of the EPR calculation. In this method, LRR can be computed using the least square regression method, where all shoreline points have been calculated in a single transect, and no shoreline points are neglected for the analysis. The sum of residual squares is minimized due to the placement of the regression line in the LRR algorithm. LRR

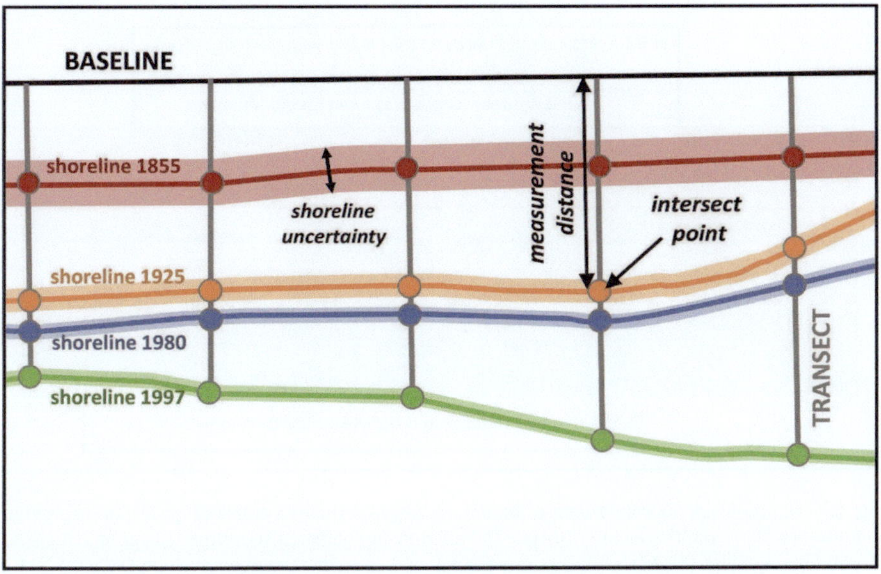

Fig. 2.1 Shoreline and baseline shapefiles with DSAS generating transects (*Source* USGS)

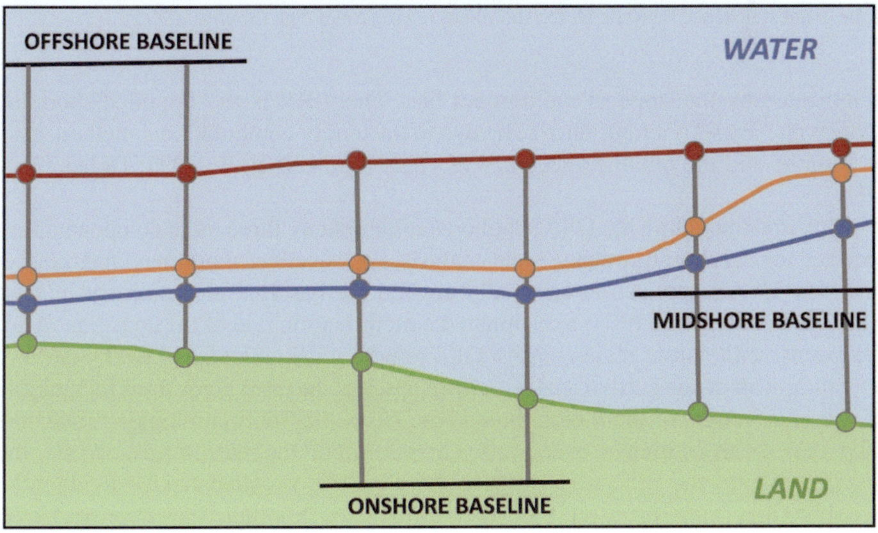

Fig. 2.2 An example of how to place a baseline onshore, offshore, or through the data (mid-shore). The resulting transects will intersect the shorelines, regardless of placement. Baselines in a dataset may be all of one type or a combination of all three

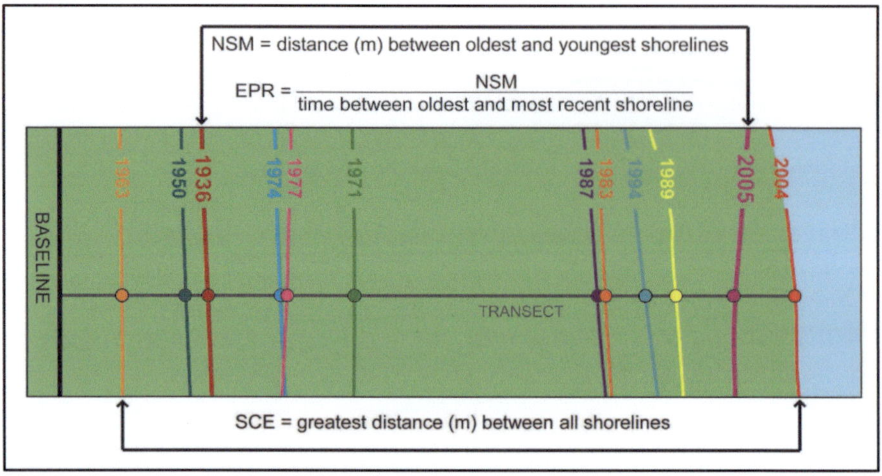

Fig. 2.3 Example: A shoreline dataset including baseline (black), transect (gray), and shoreline and intersect data (multicolor) to illustrate the relationship between shoreline change statistics: net shoreline movement (NSM), endpoint rate (EPR), and shoreline change envelope (SCE). NSM is the distance along the transect in meters (m) between the oldest shoreline (1936, red) and the most recent shoreline (2005, magenta). The EPR is the NSM distance divided by the time between the oldest (1936, red) and most recent (2005, magenta) shorelines (69 years in this example). The SCE is the greatest distance between all the shorelines regardless of date (*Source* USGS)

is indicated by the slope of the transect line. This LRR is most applied shoreline change rate analysis a long-term scale due to the purely computational method, easy to employ, and the obtained accuracy of value (Crowell et al., 1997; Dolan et al., 1991).

After understanding the DSAS tool operating system, three main components are required for the shoreline change rate analysis, i.e., baseline, shorelines, and transect line. The DSAS application especially applied the baseline measurement method (Clow & Leatherman, 1984) to estimate the modification rate of the coastline. After the extensive literature review on ArcGIS, remote sensing data, and the DSAS tool operating system, the current analysis aimed to select the three algorithms for analysis of shoreline retreat of the Indian Sundarban. Here, the Short and long-term change rates have been estimated over a 20-year scale. For the short-term analysis, the research follows the EPR model, where shoreline rate is calculated by dividing the two shorelines (most recent and oldest); more than two shorelines have been neglected in this method. Another chapter of long-term analysis has applied the LRR model, where a regression line has been plotted to obtain the change rate of the shoreline. In this method, more than two shorelines can be computed and analyzed for reaching out the retreat rate of the shoreline. On the other hand, the erosion-accretion zonation map has been generated based on the modification rate of the shoreline, applying the NSM method, where the distance between shorelines has been calculated to obtain the rate value of change (as an example, Fig. 2.4).

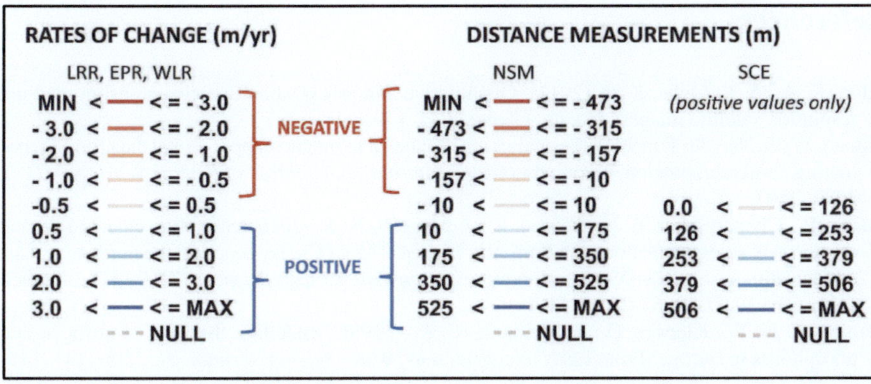

Fig. 2.4 Example, Fixed bin options for rates of change for an example dataset (left) and scaled to data (right). EPR: end point rate; LRR: linear regression rate; m, meter; m/yr, meter per year; MAX: maximum; MIN: minimum; NSM: net shoreline movement; SCE: shoreline change envelope; WLR: weighted linear regression rate

2.5 Conclusion

Nowadays, GIS and remote sensing have been widely applied tools for coastal research and development. GIS has a key functioning tool for marine, estuarine, and various hydrological studies for policy-making, sustainable planning, and conservation mitigation strategies. GIS plays a crucial role in the management of the use of coastal areas, including coastal transport, recreational extraction, waste disposal, sediment budget, erosion-accretion mapping, sustainable resource management, and research. Furthermore, GIS as a scientific application serves as a significant and adequate tool for coastal zone management, especially in providing an algorithm basis for policymakers. The development of integrated factors has led to a lack of communication between researchers and decision-makers. The spatial and temporal conditions of the coastal environment need to be analyzed and modeled through a proper data analysis method. GIS and remote sensing data, as a combined application, have been providing the most reliable and cost-effective medium of coastal research. GIS can apply the multi-dated satellite images obtained by various sensors under multi-resolution conditions. When DSAS is combined as an extension tool of ArcGIS, it provides the maximum accuracy of analysis. DSAS is a fully statistical algorithm-based data analysis method, where a researcher can obtain the shoreline change rate of any specific coastal region on both short and long-term scales. Application of the DSAS tool through the GIS platform is a comprehensive tool for erosion-accretion map generation to identify the vulnerable area due to coastal erosion. The current research work has attempted to show how remote sensing and GIS technology can be applied to the shoreline modification study over the Indian Sundarbans. This modification analysis will help a lot in understanding how the shoreline area can be managed in a sustainable way in the future and how the surrounding area can be protected from erosion and reduce mangrove loss.

References

Addo, K. A., & Kufogbe, K. S. (2011). Quantitative analysis of shoreline change using medium resolution satellite imagery in keta. *Ghana, 1*(1), 1–9.

Ahmed, W., & Neil, D. T. (1994). An evaluation of landsat thematic mapper digital data for discriminating coral reefzonation, heron reef (GPR). *International Journal of Remote Sensing, 15*(13), 2583–2597.

Anand, R., Chandrasekar, B. N., Kaliraj, S., & Magesh, N. S. (2016). Shoreline change rate and erosion risk assessment along the Trou Aux Biches—Mont Choisy beach on the northwest coast of Mauritius using GIS-DSAS technique. *Environmental Earth Sciences, 75*(5), 1–12. https://doi.org/10.1007/s12665-016-5311-4.

Beusen, A. H. W., Klepper, O., & Meinardi, C. R. (1995). Modelling the flow of nitrogen and phosphorus in Europe: From loads to coastal seas. *Water Science Technology, 31*(8), 141–145.

Blades, M. (1991). The development of the abilities required to understand spatial representations. In D. M. Mark, & A. U. Frank (Eds.), *Cognitive and linguistic aspects of geographic space: An introduction* (pp. 81–116). Dordrecht: Kluwer Academic Publishers. ISBN: 978-3-642-34358-2.

Cances, M., Font, F., & Gay, M. (2000). Principles of geographic information systems used for earth observation data. *Surveys in Geophysics, 21*(2), 187–199.

Charlier, R. H. (2000). Integrated coastal zone management and GIS, BSOC leadership seminar using today's scientific knowledge for the Black Sea areas development tomorrow. In A. S. Bologa, & R. H. Charlier (Eds.), *Proceedings* (pp. 97–108). ISBN: 973-99312-9-4.

Clow, J. B., & Leatherman, S. P. (1984). Metric mapping: An automated technique of shoreline mapping. In *44th American Congress on Surveying and Mapping. The American Congress on Surveying and Mapping, Falls Church, VA, USA* (pp. 309–318).

Cracknell, A. P., Mac, N., Fairlane, M. K., Chalton, J., & Ulbricht, K. A. (1982). Remote sensing in Scotland using data received from satellite—A study of the Tay estuary region using Landsat multispectral scanning imagery. *International Journal of Remote Sensing, 3*(2), 113–138.

Crosetto, M., Tarantola, S., & Saltelli, A. (2000). Sensitivity and uncertainty analysis in spatial modelling based on GIS. *Agricultural Ecosystem Environment, 81*(1), 71–79.

Crowell, M., Douglas, B. C., & Leatherman, S. P. (1997). On forecasting future U.S. shoreline positions—A test of algorithms. *Journal of Coastal Research, 13*(4), 1245–1255.

Cui, B. L., & Li, X. Y. (2011). Shoreline change of the Yellow River estuary and its response to the sediment and runoff (1976–2005). *Geomorphology, 127*(1–2), 32–40. https://doi.org/10.1016/j.geomorph.2010.12.001.

Dolan, R., Fenster, M. S., & Holme, S. J. (1991). Temporal analysis of shoreline recession and accretion. *Journal of Coastal Research, 7*, 723–744.

Durduran, S. S. (2010). Shoreline change assessment on water reservoirs located in the Konya Basin Area, Turkey, using multitemporal landsat imagery. *Environmental Monitoring and Assessment, 164*(1–4), 453–461.

Emran, A., Rob, M. A., Kabir, M. H., & Islam, M. (2016). Modeling spatio-temporal shoreline and areal dynamics of coastal island using geospatial technique. *Modeing Earth Systems and Environment, 2*(1), 4. https://doi.org/10.1007/s40808-015-0060-z.

ESRI. (2014). Geoenabling marine water quality monitoring in Abu Dhabi. Retrieved January 10, 2017, from http://www.esri.com/esrinews/arcnews/spring14articles/geoenablingmarine-water-quality-monitoring-in-abudhabi.

Gahegan, M., & Ehlers, M. (2000). A framework for the modelling of uncertainty between remote sensing and geographic information systems. *ISPRS Journal Photogramme Remote Sensing, 55*(3), 176–188.

Garcia, G. M., Pollard, J., & Rodriguez, R. D. (2000). Origins, management, and measurement of stress on the coast of Southern Spain. *Coastal Management, 28*(3), 215–234.

Golledge, R. G. & Stimson, R. J. (1987). *Analytic behavioural geography* (p. 355). Beckenham: Croom Helms. ISBN: 1572300493.

Goodchild, M. F. (2000). Foreword. In D. J. Wright, & D. J. Bartlett (Eds.), *Marine and coastal geographical information systems* (p. 15). London: Taylor and Francis. ISBN: 9780748408627.
Hamre, T. (1993). User requirement specification for a marine information system. Nansen Environmental and Remote Sensing Center, Technical Report 74.
Hill, G. J. E., Kelly, G. D., Tuiloma, P., & Phinn, S. R. (1994). Assessment of SPOT photographic imagery for mangrove in Figi. *Asian Pacific Remote Sensing Journal, 7*(1), 1–7.
Himmelstoss, E. A., Thierler, E. R., Zichichi, J., & Ergul, A. (2009). DSAS 4.0 installation instructions and user guide. Updated for version 4.3 (only compatible with ArcGIS 10). In U.S. Geological Survey Open-File Report 2008–1278.
Hwang, D., Karimi, H. A., & Byun, D. W. (1998). Uncertainty analysis of environmental models within GIS environments. *Computer Geoscience, 24*(2), 119–130.
Islam, A. M., & Hasan, H. (2020). Generation of IDF equation from catchment delineation using GIS. *Civil Engineering Journal, 6*(3), 540–547. https://doi.org/10.28991/cej-2020-03091490.
Kabuth, A. K., Kroon, A., & Pedersen, J. B. T. (2013). Multi decadal shoreline changes in Denmark multi decadal shoreline changes in Denmark. *Journal of Coastal Research, 30*(4), 714–728. https://doi.org/10.2112/JCOASTRES-D-13-00139.1.
Kumar, A., & Jayappa, K. S. (2009). Long and short-term shoreline changes along Mangalore coast, India. *International Journal of Environmental Research, 3*(1), 177–188.
Leatherman, S. P., & Clow, J. B. (1983). UMD shoreline mapping project. *IEE Geoscience and Remote Sensing Society Newsletter, 22*, 5–8.
Li, X., & Damen, M. C. J. (2010). Shoreline change detection with satellite remote sensing for environmental management of the Pearl River Estuary, China. *Journal of Marine Systems, 82*, S54–S61. https://doi.org/10.1016/j.jmarsys.2010.02.005.
Li, R., & Saxena, N. K. (1993). Development of an integrated marine geographic information system. *Marine Geodesy, 16*(4), 293–307.
Li, R., Keong, C. W., Ramcharan, E., Kjerfve, B., & Willis, D. (1998). A coastal GIS for shoreline monitoring and management: Case study in Malaysia. *Surveying and Land Information Systems, 58*(3), 157–166.
Li, R., Liu, J. K., & Felus, Y. (2001). Spatial modelling and analysis for shoreline change detection and coastal erosion monitoring. *Marine Geodesy, 24*(1), 1–12.
Lian, L., & Jianfei, C. (2011). Spatial-temporal change analysis of water area in Pearl River delta based on remote sensing technology. *Procedia Environmental Sciences, 10*, 2170–2175. https://doi.org/10.1016/j.proenv.2011.09.340.
Lloyd, R. (1997). *Spatial cognition: Geographic environments* (p. 221). Dordrecht: Kluwer Academic Publishers. ISBN: 978-94-017-3044-0.
Lucas, A., Abbedissen, M. B., & Budgell, W. P. (1994). A spatial metadata management system for ocean applications: Requirements analysis. In *ISPRS Working Group II/2 Workshop on the Requirements for Integrated GIS* (*New Orleans, Louisiana*) (pp. 1–13).
MacDonald, A., & Cain, M. (2000). Marine environmental high risk areas (MEHRAs) for the UK. *International Maritime Technology, 112*(2), 61–70.
Marghany, M., Sabu, Z., & Hashim, M. (2010). Mapping coastal geomorphology changes using synthetic aperture radar data. *International Journal of Physical Sciences, 5*(12), 1890–1896.
Masria, A., Nadaoka, K., Negm, A., & Iskander, M. (2015). Detection of shoreline and land cover changes around Rosetta promontory, Egypt, based on remote sensing analysis. *Land, 4*(1), 216–230. https://doi.org/10.3390/land4010216.
Maurya, D. M., Thakkar, M. G., Patidar, A. K., Bhandari, S., Goyal, B., & Chamyal, L. S. (2010). Late quaternary geomorphic evolution of the coastal zone of Kachchh, Western India. *Journal of Coastal Research, 24*(3), 746–758.
Maurya, D. M., Tiwari, P., Shaikh, M., Patidar, A. K., Vanik, N., Padmalal, A., & Chamyal, L. S. (2021). Late quaternary drainage reorganization assisted by surface faulting: The example of the Katrol Hill Fault zone, Kachchh, Western India. *Earth Surface Processes and Landforms, 46*(7), 1268–1293. https://doi.org/10.1002/esp.5097.

Mujabar, P. S., & Chandrasekar, N. (2011). Shoreline change analysis along the coast between Kanyakumari and Tuticorin of India using remote sensing and GIS. *Arabian Journal of Geosciences*. https://doi.org/10.1007/s12517-011-0394-4.

Murali, R. M., Dhiman, R., Choudhary, R., Seelam, J. K., Ilangovan, D., & Vethamony, P. (2015). Decadal shoreline assessment using remote sensing along the central Odisha coast, India. *Environmental Earth Sciences, 74*(10), 7201–7213. https://doi.org/10.1007/s12665-015-4698-7.

Nandi, S., Ghosh, M., Kundu, A., Datta, D., & Baksi, M. (2016). Shoreline shifting and its prediction using remote sensing and GIS techniques: A case study of Sagar Island, West Bengal (India). *Journal of Coastal Conservation, 20*, 61–80. https://doi.org/10.1007/s11852-015-04184.

Nassar, K., Mahmod, E. W., Fath, H., Masria, A., Nadaoka, K., & Negm, A. (2018). Shoreline change detection using DSAS technique: Case of North Sinai Egypt. *Marine Geo Resources & Geotechnology*. https://doi.org/10.1080/1064119X.2018.1448912.

Nayak, S. R. (2002). Use of satellite data in coastal mapping. *Indian Cartographer, 22*, 147–157.

Nayak, S. R., & Sahai, B. (1985). Coastal morphology: A case study of the gulf of Khambhat (Cambay). *International Journal of Remote Sensing, 6*(3–4), 559–567.

Nierenberg, W. A. (1992). *Encyclopedia of earth system science* (p. 2825). San Diego, California: Academic Press. ISBN: 978-0122267192.

Nyerges, T. L. (1991). Analytical map use, cartography and geographic information system. *Cartography and Geographic Information Systems, 18*(1), 22–28.

Psuty, N. P., Steinberg, P. E., & Wright, D. J. (2002). Coastal and marine geography. In G. L. Gaile, & C. J. Willmott (Eds.), *Geography in America at the dawn of the 21st century* (p. 333). New York: Oxford University Press. ISBN: 9780199295869.

Ramalingam, G., & Renganathan, M. (1998). Coastal geomorphology and its evolution—A case study in Agniar basin Tamilnadu. *Journal of Indian Geomorphology, 3*(2), 146–156.

Ryu, J. H., Won, J. S., & Min, K. D. (2002). Waterline extraction from Landsat TM data in a tidal flat: A case study in Gosmo Bay, Korea. *Remote Sensing of Environment, 83*(3), 442–456. https://doi.org/10.1016/S0034-4257(02)00059-7

Saranathan, E., Chandrasekaran, R., Soosai Manickaraj, D., & Kannan, M. (2011). Shoreline changes in Tharangampadi village, Nagapattinam district, Tamil Nadu, India—A case study. *Journal of the Indian Society of Remote Sensing, 39*(1), 107–115. https://doi.org/10.1007/s12524-010-0052-4.

Siddiqui, M. N., & Maajid, S. (2004). Monitoring of geomorphological changes for planning reclamation work in coastal area of Karachi, Pakistan. *Advances in Space Research, 33*(7), 1200.

Singh, A. (1989). Digital change detection techniques using remotely sensed data. *International Journal of Remote Sensing, 10*(6), 989–1003. https://doi.org/10.1080/01431168908903939.

Slater, F. (1982). *Learning through geography* (p. 340). Oxford: Heineman Educational Books. ISBN: 0435354108.

Suresh, P. K., & Sundar, V. (2011). Comparison between measured and simulated shoreline changes near the tip of Indian Peninsula. *Journal of Hydro-Environment Research, 5*(3), 157–167. https://doi.org/10.1016/j.jher.2011.02.002.

Tran Thi, V., Tien Thi Xuan, A., Phan Nguyen, H., Dahdouh- Guebas, F., & Koedam, N. (2014). Application of remote sensing and GIS for detection of long-term mangrove shoreline changes in Mui Ca Mau Vietnam. *Biogeosciences, 11*(14), 3781–3795. https://doi.org/10.5194/bg-11-3781-2014.

Valavanis, V. D. (2002). *Geographic information systems in oceanography and fisheries* (p. 173). New York: Taylor and Francis Inc. ISBN: 9780415284639.

Van Zuidam, R. A., Farifteh, J., Eleveld, M. A., & Cheng, T. (1998). Developments in remote sensing, dynamic modelling and GIS applications for integrated coastal zone management. *Journal Coastal Conservation, 4*(2), 191–202.

Wagner, T. W., Michalek, J. L., & Laurin, R. (1991). Remote sensing application in the coastal zone: A case from the Dominican Republic. In R. W. Stoffle & D. D. Halmo (Eds.), *International earth science information network report* (pp. 17–56). University Center.

Wang, X. Z., Zhang, H. G., Fu, B., & Shi, A. (2013). Analysis on the coastline change and erosion-accretion evolution of the Pearl River estuary, China, based on remote-sensing images and nautical charts. *Journal of Applied Remote Sensing, 7*(1), Article 073519. https://doi.org/10.1117/1.JRS.7.073519.

White, K., & El Asmar, H. M. (1999). Monitoring changing position of coastlines using thematic mapper imagery, an example from the Nile Delta. *Geomorphology, 29*(1–2), 93–105. https://doi.org/10.1016/S0169-555X(99)00008-2.

Williamson, P. (1994). Integrating earth system science. *Ambio, 23*(1), 3. ISSN: 0044-7447.

Yamano, H., Shimazaki, H., Matsunaga, T., Ishoda, A., McClennen, C., Yokoki, H., et al. (2006). Evaluation of various satellite sensors for waterline extraction in a coral reef environment: Majuro Atoll, Marshall Islands. *Geomorphology, 82*(3–4), 398–411. https://doi.org/10.1016/j.geomorph.2006.06.003.

Yanli, T. (2011). *The application of GIS and RS for coastline change detection and risk assessment to enhanced sea level rise*. International Institute for Geo Information Science and Earth Observation.

Chapter 3
Exploring Short-Term Shoreline Dynamics in the Indian Sundarbans: Digital Shoreline Analysis System (DSAS) Approach

Abstract Geospatial analysis is the most crucial application for assessing the shoreline modification rate on a short-term scale. The present chapter has been exploring the application of GIS for estimating the short-term modification rate of the shoreline along part of the Indian Sundarban. A geospatial tool combined with various quantitative and statistical algorithm-based models is more accurate, reliable, and cost-effective to understand the erosion-accretion rate of the shoreline. To identify the accurate change rate, the chapter follows the Digital Shoreline Analysis System (DSAS) tool, which is an extension of the ArcGIS application. The present chapter identifies the modification rate over the Indian Sundarbans on a short-term scale during 2000–2020, which is also divided into two segments, i.e., 2000–2010 and 2010–2020. Four littoral zones have been generated to complete the analysis with the selected study area. The histogram threshold model has been used here to extract the actual shoreline position using Landsat 8 and Landsat TM satellite images with 30 m resolution. In DSAS, the End Point Rate (EPR) has been applied to calculate the short-term change over the region. A brief short-term change analysis of the last 20 years reveals some drastic changes in the region. In the case of segment 1, the first short-term (2000–2010) analysis shows that there is a lot more area in the low accretion zone, but later it becomes shifted toward the low erosional zone in 2010–2020. The most significant changes have been identified in Bangaduni and Sagar Island. In the case of these two islands, the analysis of the first short-term period found a large number of accretion zones, but later the entire islands changed from a low erosion to a moderate erosion zone. Increased erosion has had a significant impact on the mangroves, as well as the saltwater intrusion over the region, which has also been observed.

Keywords Short-term Analysis · Shoreline Changes Dynamics · Digital Shoreline Analysis System (DSAS) · End Point Rate (EPR) · Littoral Zones · Satellite Images · Uncertainty · Geospatial Techniques · Indian Sundarbans

© The Author(s), under exclusive license to Springer Nature Switzerland AG 2026

A. Nath et al., *Shoreline Modification Over the Indian Sundarban*, SpringerBriefs in GIS, https://doi.org/10.1007/978-3-032-13981-8_3

3.1 Introduction

Shoreline or strandline is very dynamic in characteristics in both spatial and temporal aspects, and the change systems are responsible for the fluctuation factors of tides, storm surges, sea level rise, and various geotectonic movements in the upper and lower earth systems (Woodworth et al., 2019). These changes may be short-term or long-term, caused by several geomorphological and hydrological as well as endogenic activities or some suddenly generated short-term agents on the Earth's surface (Karunarathna et al., 2018). Shoreline position is directly affected by the sea level rise, especially by the damage to coastal ecosystems and erosion, such as tidal flats and mangroves (Thakur et al., 2018). Recently, global sea level acceleration has been a major concern, as showcased in different research models and frameworks (Kulp & Strauss, 2019). The low-lying coastal belts are mostly vulnerable due to the sea level rise. The most common problem in any coastal region is erosion-accretion, where the landward movement of the shoreline is known as erosion, whereas the seaward movement is known as accretion or sedimentation. Erosional work is directly responsible for land degradation, submergence of land, loss of dunes, saltwater intrusion in agricultural fields, and loss of mangrove vegetation (Williams et al., 2018). Mangrove distribution in the coastal region is adversely affected by shoreline shifting and sea level changes (Ward et al., 2016).

Shoreline has been modified in spatial and temporal ways, mainly caused by the attrition or accretion changes of the dunes. This complex process is made by the several interactions of coastal morphology with wave energy that reaches the shoreline, sea level rise, dune degradation, and sediment transportation (Vaz & Bowman, 2013). Higher sedimentation gave the result of accretion, whereas extreme sediment transportation provides the result of coastal zone regression (Narayana, 2016). The man-made activities are also caused by the shoreline attrition, such as sand mining, fishing, and urban structures near the shoreline. The accrual or attrition has been creating immense pressure on coastal mangrove vegetation to sustain the coastal ecosystem.

The tidal and alkaline water interaction zones have been enriched by some typical ecological groups of trees and shrubs that are known as mangrove vegetation, mainly found in tropical coastal regions. Mostly low-energy tidal-influenced areas and domination of sedimentation are ideal areas for mangrove growth. The mangroves are preferably increased by the fixed range of tide and low to moderate elevation, a perfect balance of salinity and availability of groundwater, and a frequently inundated area for proper growth of this vegetation (Ramon et al., 2019). It has been recorded in Alongi's (2015) report that 70% of mangrove decline is due to several anthropogenic activities and climate change issues. This enhancement of mangrove decline has also accelerated the global warming, sea level rise, storm surges, and coastal erosion that heat up the coastal ecosystem as well as long-term loss of mangrove health (Ding, 2016; Kassakian et al., 2017).

The shoreline change is creating a direct effect on the mangrove zone, where the erosion of the shoreline has occurred, and there is also a decline of mangroves.

In deltaic areas, it has been observed that due to the increase in erosion rate, the mangrove shifted toward the landward side or degraded because of bank slumping (Chaudhuri et al., 2019). On the other hand, mangrove sediment salinity has increased due to shoreline erosion, which leads to groundwater depletion. After understanding the scarcity of mangrove sustainability, it has become imperative to reduce the coastal shoreline modification in a sustainable process and mitigate the erosion to some extent. The proper planning of the coastal zone to protect the coastal ecosystem, the precise prediction of shoreline change can be helpful to land subsidence or coastal erosion to reduce the threats (Passeri et al., 2015). Short-term and long-term changes in shoreline position analysis are a significant way to understand the future migration of the shoreline.

To identify the shoreline migration, several conventional methods are applied to compare the past relative data, i.e., Linear Regression Rate (LRR), Average of Rates (AOR), End Point Rate (EPR), and Jackknife (JK) (Cham et al., 2019; Dolan et al., 1991; Fenster et al., 1993; Gholami & Baharlouii, 2019; Jana et al., 2016). In the current research context, geographical information systems (GIS) and remote sensing (RS) technology have been widely applied to assess the shoreline modification and erosion-accretion zonation mapping (Jayappa et al., 2006; Ryu et al., 2002; Siddiqui & Maajid, 2004; Singh, 1989; Yamano et al., 2006). To obtain reliable accuracy, the researchers mostly depend on the GIS method for monitoring the coastal changes (Emran et al., 2016; Kona et al., 2025; Valderrama-Landeros et al., 2024; Wang et al., 2013; Yanli, 2011). Several neo-tectonic GIS methods have also been applied for assessing the river basin or catchment area modification (Argyriou et al., 2017; Gupta et al., 2022; Islam & Hasan, 2020; Maurya et al., 2010, 2021; Taib et al., 2024).

The best part of GIS technique is, it can be assessing the multi-resolution spatial temporal satellite images with maximum resolution coverage to shoreline modification analysis that also applied in globally by several researchers at different time (Carter & Schubert, 1974; Wagner et al., 1991; Ahmed & Neil, 1994; Cracknell et al., 1982; Nayak & Sahai, 1985; Anbarasu et al., 1999; Ramalingam & Renganathan, 1998; Nayak, 2002; Hill et al., 1994; Kabuth et al., 2013; Addo & Kufogbe, 2011; Murali et al., 2015; Tran Thi et al., 2014; Anand et al., 2016; Kermani et al., 2016; Masria et al., 2015; Nandi et al., 2016; Nassar et al., 2018; Suresh & Sundar, 2011; Nath et al., 2023, 2024a, b). Mujabar and Chandrasekar (2011) show the accuracy of multi-resolution satellite data for obtaining the shoreline changes in periodic assessments. Saranathan et al. (2011) also stated that remote sensing data provides more precise accuracy to shoreline detection analysis than the conventional method. Remote sensing is a more significant method to detect and evaluate the coastline position using multi-resolution satellite data, reported by Marghany et al. (2010).

After an extensive literature review of shoreline change analysis, the current research work aimed to identify the shoreline position on a short-term basis from 2000 to 2020 and also assess the change rate of the shoreline in the Indian Sundarbans. The study has also aimed to mark the erosion-accretion zones over the region using the transect method through GIS and remote sensing. The current study follows the Digital Shoreline Analysis System (DSAS) using the most reliable method of

End Point Rate (EPR), which has also been used by several researchers in the recent past (Jana et al., 2013; Mansour et al., 2024; Mukhopadhyay et al., 2012; Nandi et al., 2016; Nassar et al., 2018; Nath et al., 2024a; Ramachandran et al., 2025). The entire study (Indian Sundarbans part) area has been divided into four segments to assess the short-term shoreline modification precisely.

This chapter investigates the following objectives: (i) to identify the temporal shoreline positions; (ii) to identify the shoreline migration rate and analyze the succession and regression of shoreline using the DSAS model; (iii) to identify the short-term erosion-accretion zone of the part of the Indian Sundarban. To acquire the objective, study using the satellite image and several statistical analyses in the GIS platform.

From past studies, it has been observed that land reclamation and land abrasion are ongoing processes of the Sundarbans in a simultaneous way (Chatterjee et al., 2015). The past study by Bera and Maiti (2019) revealed that most of the Indian Sundarbans Islands are under high threats of coastal erosion using the DSAS tool. Another study by Hale et al. (2019) remarked that the lack of sediment transport has caused the retreating delta formation at Sundarbans. The mangrove vegetation is found under immense pressure due to shoreline change (Sahana et al., 2019). Another study has also been conducted by Thakur et al. (2017) at Namkhana Bakkhali Island about shoreline modification and the impact of land use and land cover change. In that study, it was observed that most part of the islands suffered from coastal regression, and few amounts of accretion remarked. Tripathi et al. (2018) have also reported that sea level rise caused by the isostatic sinking and tilting of the delta. A study by Thakur et al. (2021) on land use and land cover change using NDVI revealed that mangrove area decreased due to land retreat and dune degradation. Due to all this literature survey, a complete research work is required for obtaining the selected objectives. The current chapter has evolved to find the answer to the above-mentioned objectives.

3.2 Methodology

3.2.1 Data Use and Image Processing

The shoreline modification study was implemented using three cloud-free multi-resolution Landsat images of 2000, 2010, and 2020 that were collected from the USGS site. Table 3.1 shows the details of the applied satellite images. As the modification of the shoreline is a very slow physical process, a minimum of 10 years' data is required for precise analysis in case of short-term change assessment. A total of 20 years has been calculated to assess the shoreline modification. The total of 20 years has been divided into two periods, 2000–2010 and 2010–2020, to calculate the short-term change assessment over the study region. In the case of shoreline

Table 3.1 Applied data for the present study

Year	Sensor	Resolution (m)	Path/row	Date	Projection	Datum	Source
2000	Landsat TM	30	138/45	11/9/2000	UTM 45	WGS84	USGS
2010	Landsat TM	30	138/45	12/23/2010	UTM 45	WGS84	USGS
2020	OLI/TIRS	30	138/45	12/18/2020	UTM 45	WGS84	USGS

shifting research work, satellite data provides the maximum reliability and accuracy (Hagenaars et al., 2018; Esmail et al., 2019).

After collecting the required satellite images, they were processed through the ArcGIS platform. From past coastal studies, it has been observed that Landsat Thematic Mapper and Landsat OLI 8 series images are more reliable and suitable for shoreline shifting study (Emran et al., 2016; Jana et al., 2016; Kaliraj et al., 2015; Maiti & Bhattacharya, 2009; Nandi et al., 2016). The recent shoreline modification study has also applied the Landsat Thematic Mapper and Landsat 8 series satellite images to obtain the true shoreline position. To extract the shoreline, all images have been projected with the WGS 84 datum as a UTM projected system (Universal Transverse Mercator) 45 zone. The false color composite method has been applied to extract the shoreline from the images.

3.2.2 Shoreline Extraction Process

The automatic extraction method for shoreline is quite complicated due to the high-water saturation in the land–water boundary for extracting the shoreline (Hossain et al., 2021; Jana et al., 2013; Nandi et al., 2016; Nassar et al., 2018; Tsai, 2024). The current research follows the histogram threshold method to extract the true coastline position, it also known as the band ratio model, a method followed by Maiti and Bhattacharya (2009). Two band ratios have been applied in this method as a binary image, i.e., Band 2/Band 5 $\geq$ 1.0 and Band 2/Band 4 $\geq$ 1.0. Reflectance of ocean water is to be considered for extracting the coastline, where a value close to zero is found in the infrared band in the case of single-band satellite data. Higher than zero is considered land value in this model. Applying the model, all water pixels remarked as water features, and land features were obtained from land pixels. The entire shoreline extraction method has been portrayed by a flow diagram for better understanding (Fig. 3.1).

3.2.3 Shoreline Position Validation

The band ratio or histogram threshold method is one of the most applied methods to validate the true shoreline position. The exact shoreline position can be acquired by

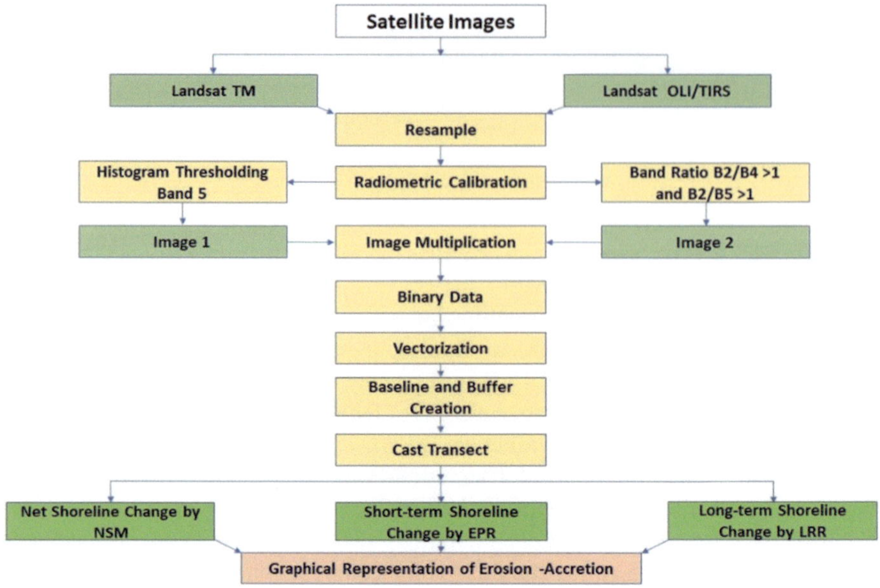

Fig. 3.1 Flow diagram of the applied model

this method. From single-band satellite data, the shoreline has been extracted based on the ocean water reflectance that is close to or equal to zero. Land features have been remarked as more than zero in this method. The most appropriate band is band 5, which is mid infrared, to acquire the shoreline from the land–water boundary in the case of Landsat Thematic Mapper (Kelley et al., 1998). The band 5 (mid infrared) has very low reflectance that is easily absorbed by the water, whereas land features like soil and vegetation provide high reflectance among the other bands. By this method, water and land features are clearly identified by the researchers. To classify the water and land features, the threshold value is very significant. By this band ratio model, all water features fall under the water value and land features fall under the land pixel value, where one value remarked as a water pixel, and land pixels are considered as zero value. After that, image 1 has been obtained, which is a binary image. For reaching out the image 2, the second process has been applied on Band 2/Band 4 and Band 2/Band 5. From the second method, water and land pixels are clearly separated. Similarly, one-pixel value remarked as water, and zero shows the land pixels. After deriving both images, multiply the images to avoid the errors in the final image. After multiplying, a final binary image has been achieved. From the final image, the shoreline has been extracted manually. Here, Google Earth Explorer has been used to validate the historical coastline, and the 2020 image was validated by ground truth data through Global Positioning System (GPS).

3.2.4 Shoreline Modification Rate Calculation

The current work has been completed using the DSAS analysis method through the ArcGIS platform to calculate the shoreline modification rate (Thieler et al., 2005). In this calculating system, a baseline has been generated from the buffer distance of the shoreline, and transects were drawn from the baseline at a perpendicular angle toward onshore and offshore areas of the coastal zone. This transect distance was calculated as a periodical change of shoreline (Maiti & Bhattacharya, 2009; Muskananfola & Febrianto, 2020). From this method, maximum accuracy has been obtained, so this is the most applied method for calculating the shoreline boundary. For this analysis, the required parameters are (i) baseline along with the essential attributes, (ii) generating the buffer from baseline, and iii) reaching out to the coastline position with the necessary attributes.

The present study has chosen the maximum part of the Indian Sundarbans to identify the modification of the shoreline. Due to the large area covered in the research work, the area was divided into four segments, mostly the littoral zone. The segments are divided into a littoral zone; the zones are:

(i) **Segment 1** is from Ramtanunagar to Sundarbans National Park,
(ii) **Segment 2** is covered by Bangaduni Island,
(iii) **Segment 3** covered by Sagar Island and
(iv) **Segment 4** is covered by Murmu Island.

Segment 1 is covered by Minnon Island, Mousuni Island, Bakkhali, Lalganj Island, Kalas Beach, Jharkhali, etc.

These four segments have been divided by the littoral zone selection process. Bray et al. (1995) revealed in their research that mostly similar sediment conditions and similar direction of sediment transport have been considered for choosing the littoral zone (Deepika et al., 2014; Jana et al., 2016; Pranzini et al., 2020). This identification process is significant to understand the shoreline modification on a regional scale.

Through the DSAS analysis process, transect lines have been generated from the baseline at a perpendicular angle for all segments (littoral zones). Various statistical methods like End Point Rate (EPR) and Net Shoreline Movement (NSM) have been applied mostly for short-term behavioral changes analysis (Jana et al., 2013; Das et al., 2013; Thieler et al., 2019; Himmelstoss et al., 2021; Nath et al., 2020, 2024a). But the EPR model is mostly applied for calculating the short-term change rate in any coastal areas from the recent past (Dolan et al., 1991; Hwang et al., 2014; Thieler et al., 2005; Ayadi et al., 2016; Nassar et al., 2018). The net distance of two successive shorelines has been calculated in the NSM method, where the rate is not considered. But on the other side, EPR has calculated the change or alteration rate between only two shorelines in m/year. In the EPR model, the youngest shoreline has been divided by the oldest shoreline to reaching out the rate of change with time. The same process has been applied for NSM, but only the distance of the shoreline can be obtained. Each transect value has shown the erosion-accretion rate of the shoreline, where the landward movement of the transect represents the erosion, and seaward movement remarked as accretion of the shoreline. EPR has been calculated by Eq. (3.1).

$$EPR = L1 - L2/t1 - t2 \quad (3.1)$$

Here, $L1$ represents the youngest shoreline, and $L2$ represents the oldest one from the baseline. On the other side $t1$ and $t2$ demarked the corresponding years of shorelines.

3.2.5 Calculation of the Uncertainty of Shoreline Data

Application of various satellite data with several time periods of shoreline obtaining requires quantifying the uncertainties and errors of the data. Various studies have been providing and addressing insightful methods to reduce these errors and data manipulation during analysis (Kumar et al., 2024; Sánchez-García et al., 2020). Mainly, two types of uncertainty have been observed as per Romine and Fletcher (2013), the first one is positional errors that indicate the tidal fluctuations and seasonal changes; the other one is measurement errors, which occurred during rectification, digitization, and pixel correction. In this research, different types of uncertainty have been illustrated:

Tidal fluctuating error (Et): Due to the tidal fluctuation, the horizontal unpredictability of shoreline position has been indicated by the tidal error (Fletcher et al., 2012; Castelle et al., 2021; Konstantinou et al., 2023). Rise and fall in sea level are the main causes of tidal fluctuation that generate the displacement of the shoreline. In the recent study, the high tide range was used for extracting the shoreline's true position.

Seasonal error (Es): This error occurred mostly due to the domination of waves, storms, and coastal functions over the year, and due to these influences, the seasonal shoreline has been noticed naturally (Fletcher et al., 2012). To avoid this error, the current work selects all cloud-free satellite data for analysis.

Digitization error (Ed): This type of error occurred during the digitization period mostly when the Landsat data was used for analysis (Fletcher et al., 2012). Here, Landsat satellite data with 15 m spatial resolution has been obtained, where half of the pixels fall under an error of 7.5 m. This means an error can occur within half of a pixel during shoreline digitization of Landsat data. So, the shoreline has been digitized with an accurate position to a logical degree.

Rectified error (Er): The true position of the shoreline and the earth surface position has been calculated in a deviation method considered as Rectified error for the shoreline extraction process in the rectified image. This type of error is mainly generated by the error of orbital determination of satellite data, displacement of sensor geometric model, and limited ground truth points while registering the image (Fletcher et al., 2012). One-fourth of a pixel error has been found while rectifying the Landsat level 2 data, which correlated to a near 7.5 m in the current research work.

Table 3.2 Estimate the positional errors (*Note* NA indicates "not applicable")

Total position errors (m)	Selected shoreline time period		
	2000	2010	2020
Tidal fluctuating error (Et)	2.12	NA	NA
Seasonal error (Es)	NA	NA	NA
Digitization error (Ed)	2	2	3
Rectified error (Er)	NA	NA	NA
Pixel error (Ep)	6	5	3

Pixel error (Ep): This type of error is regulated by the spatial resolution of satellite data that remarked as pixel size error. Here, all the used satellite data have the same spatial resolution of 30 m that helps to avoid the pixel size error during rectification and digitization.

Total positional errors estimation: A square root method has been applied to estimate the true positional error. By this method, various errors have been defined through the square sum of all errors in the square root method (Mishra et al., 2019, 2023; Paul et al., 2024; Das et al., 2021; Fletcher et al., 2012). The true position calculation has been shown by Eq. (3.2) and Table 3.2. In the DSAS method, uncertainty has been calculated by assigning the weight of each shoreline of the Weighted Linear Regression Rate (WLR). The change has been calculated in the study through the EPR method.

$$U_t = \pm\sqrt{\left(E_t^2 + E_s^2 + E_d^2 + E_r^2 + E_p^2\right.} \quad (3.2)$$

3.3 Result and Discussion

3.3.1 Short-Term Shoreline Modification Condition of Segment 1 or Littoral Zone 1

Two short-term modification analyses have been performed to estimate the change rate over 20 years. The total 598.64 km length of shoreline has been analyzed for segment 1, divided into two periods, i.e., 2000–2010 and 2010–2020. After calculating the shoreline change rate, the erosion–accretion rate was determined using the EPR method. In this approach, negative EPR values indicate erosion (landward shoreline movement), while positive values indicate accretion (seaward shoreline movement).

Five erosion-accretion classes have been selected to show the erosion-accretion zone prominently, i.e., high erosion, moderate erosion, low erosion, low accretion, and low accretion zone. The entire segment 1 between 2000 and 2010 has been found under a mostly accreted zone, a high-to-low accretion zone. To calculate the change rate, a total of 3942 transects were generated 2000 m from baseline to seaward to cover the maximum perpendicular distance for the 2000–2010 time period. In the case of the first time period (2000–2010), only 403 transects fell under the erosional zone. The high rate of accretion has been represented by 128 transects. Near Kalas Beach, Gobardhanpur Beach, Baliara, and a small part of the Sundarbans National park areas were observed under a high accretion zone (Fig. 3.2). The maximum high accretion value was obtained near Kalas Beach of 205.67 m, which means 20.57 m/year high accretion observed in this area. But most of the region has been found under low accretion, where the values vary between 15 and 50 m accretion. High rate of erosion has been found as a value of −84.39 m, where −8.44 m/year rate of erosion was identified. Most of the low erosion rate has varied between −2 and −25 m EPR value for the first time duration. Minnon Island has been remarked as a low to moderate erosion zone with an erosion rate of 48.45 m. On the other hand, Mousuni Island, Muriganga, Lalganj Beach, Lothian Beach, Sitarampur Beach, and Sundarbans Delta have been observed as low erosional zones between 2000 and 2010 time period. The graph shows the change rate by the EPR model to represent the erosion-accretion over 2000–2010 (Fig. 3.3).

But the segment 1 has been prominently changed by the next 10-year time period (2010–2020). A clear modification has been observed during this time period. A total of 4601 transects were generated with a perpendicular distance of 2000 m from baseline, along 500 m of buffer to calculate the modification or alteration rate of the shoreline. A significant change has been marked over the segment or littoral zone

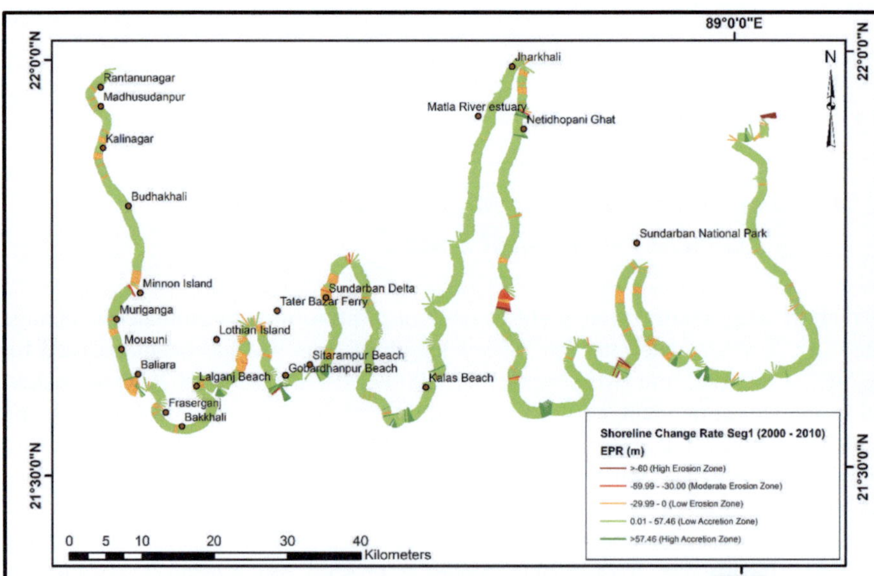

Fig. 3.2 Erosion-accretion zone of segment 1 (2000–2010)

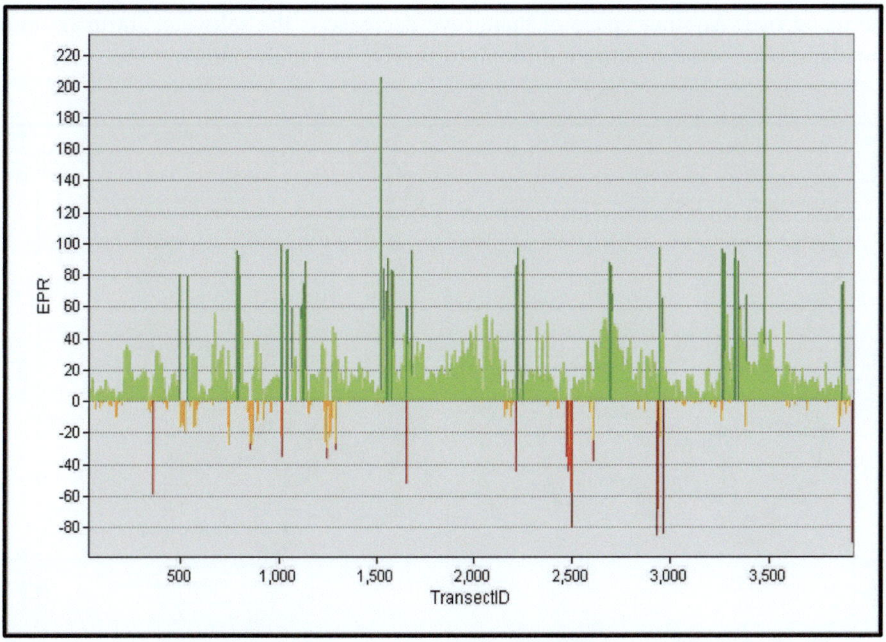

Fig. 3.3 Graph of Erosion-accretion rate by EPR (2000–2010), segment 1

where a prominent erosion enhancement has been noticed. Most of the beaches that were found under a low accretion zone in the past 10 years, shifted to a low erosional zone within the next 10 years (2010–2020). Major erosion was identified in Netidhopani Ghat and near Gobardhanpur Beach with the rate of −195.44 m (−19.54 m/year) and −166.34 m (−16.34 m/year). A total of 1174 transects represented the erosional movement of the shoreline where the negative value of EPR was obtained. Kalinagar, Mousuni, Baliara, Bakkhali, Lalganj Beach, Lothian Beach, Tater Bazar Ferry, Kalas Beach, Jharkhali, and the extension part of the Sundarbans National Park area have been perceived as having a low erosional regime (Fig. 3.4). But moderate erosion has been noticed in a very small part near Kalas Beach, with a rate of 83.25 m in EPR. The low erosion rate has varied between 10.62 m (1.06 m/year) and 51.87 m (5.18 m/year). From the obtained values, a graph has been prepared to showcase the prominent change of segment 1 between 2010 and 2020 time period (Fig. 3.5). The beaches of low erosion were found as low accretion in the past 10 years. The shoreline erosion has also been creating pressure on mangrove vegetation over the littoral zone. 6.77% mangrove area has been increased in Bakkhali Namkhana Island since 2010, but after that, the shoreline regression toward the land was the cause of mangrove reduction over the same region (Thakur et al., 2018). Mangrove loss was also observed near Frasergani due to coastal erosion. Another study by Paul et al. (2017) also reported that the necessary mud and clay have deteriorated due to the submergence of the sand bank, which reduces the seed propagation to regenerate the mangrove. Another cause of mangrove decrease is the saltwater intrusion into the mangrove forest, making it difficult to survive in a high-salinity zone.

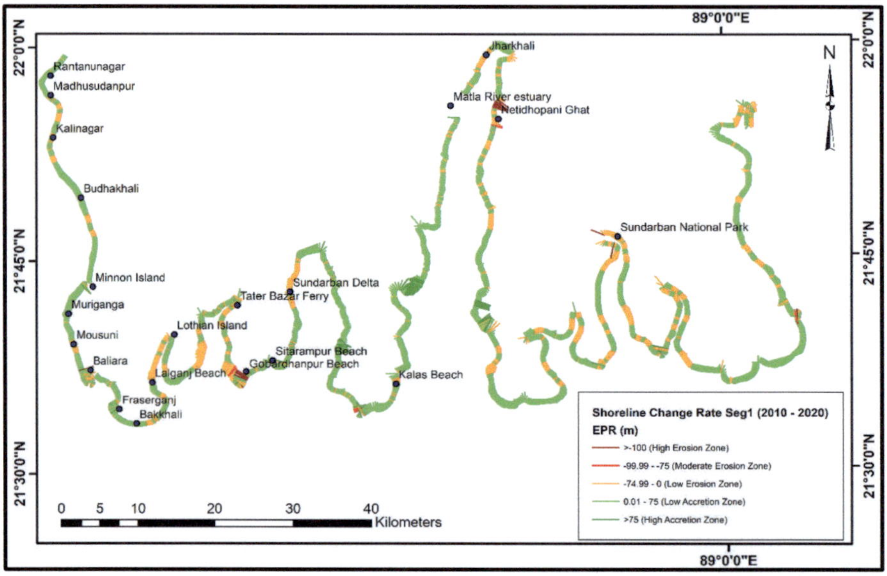

Fig. 3.4 Erosion-accretion zone of segment 1 (2010–2020)

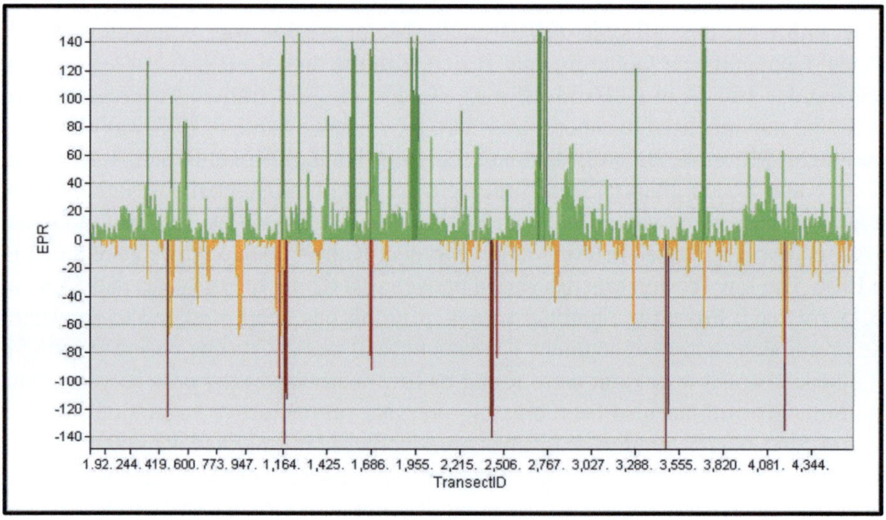

Fig. 3.5 Graph of Erosion-accretion rate by EPR (2010–2020), segment 1

3.3.2 Short-Term Shoreline Modification Condition of Segment 2 or Littoral Zone 2

The recession (erosion) and Progradation (accretion) have been observed in all shoreline positions during both periods in segment 2 named Bangaduni Island, with a length of 22.86 km of shoreline. Mostly the shoreline dynamics has been dominated by three important factors, i.e., tectonic and compaction subsidence, sea level rise and wave movement, and sediment transportation by the river and estuaries. The shoreline modification of segment 2 has been carried out by two successive time periods 2000–2010 and 2010–2020. In the first short-term analysis, it has been observed that 276 transects were applied to estimate the regression and Progradation rate. The transect distance was taken 3000 m from baseline with a 1000 m buffer area and with 50 m spacing between transects. Figure 3.6 represents the erosion-accretion zone of Bangaduni Island (2000–2010). Figure 3.7 showcases the graphical representation of shoreline alteration rate obtained by EPR (m) based on the transect ID from the baseline. Here, only 33-number transect has shown the negative shift of the shoreline, which means the landward movement. Most of the area was found under a low accretion zone value varying between 9.08 m and 62.88 m in EPR. The upper part of the Island has been receiving the highest rate of erosion or a negative trend of analysis, where −121.76 m (12.18 m/year) erosion rate was obtained by the EPR. In the lowest part of Island has also been received the high accretion value that 191.07 m (19.11 m/year) by EPR analysis. The little portion of the upper part has also been observed under a moderate to low erosional trend, where the value varies between −3.78 and −56.73 m.

On the other side, the next 10 years of the time period have shown some changes in the segment 2 shoreline position. In the next 10 years' analysis (2010–2020), 252 transects with a perpendicular distance of 3000 m from the baseline were used in the EPR method. A total of 47 transects have represented the negative trend of shoreline modification. During this time period (2010–2020), some negative changes in the shoreline have been revealed, showing landward movement. The EPR analysis of 2010–2020 has reflected some major changes within the Bangaduni Island (Fig. 3.8). The graphical representation has clearly shown the alteration rate of shoreline EPR value from the baseline (Fig. 3.9). The areas of the Island that were under low accretion have been shifted to a high accretion zone. The western part of the study area has been found the maximum erosion value of 102.3 m (10.23 m/year), shown by transect number 2. The low erosion to moderate erosion rate of the study area varies between −2.69 and −54.12 m. The low to high accretion rate varies between 12.43 and 53.99 m.

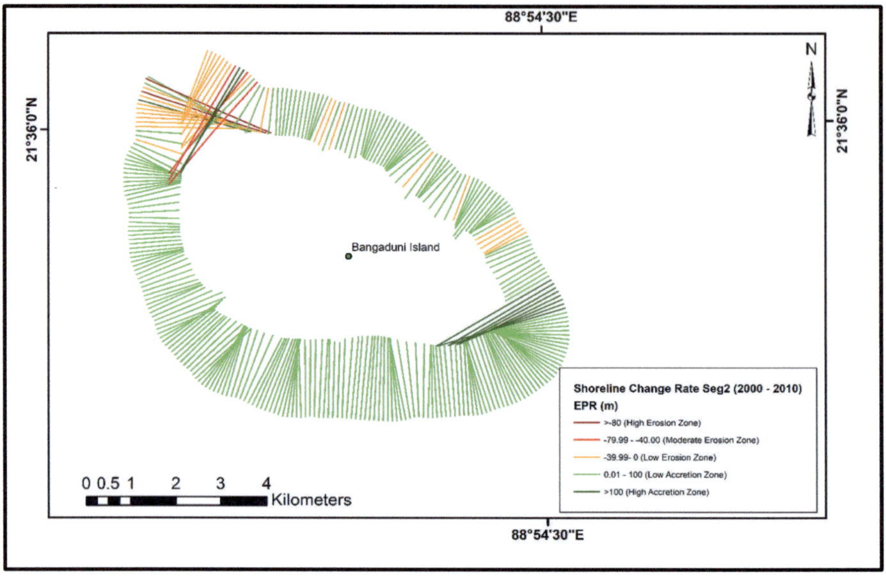

Fig. 3.6 Erosion-accretion zone of segment 2 (2000–2010)

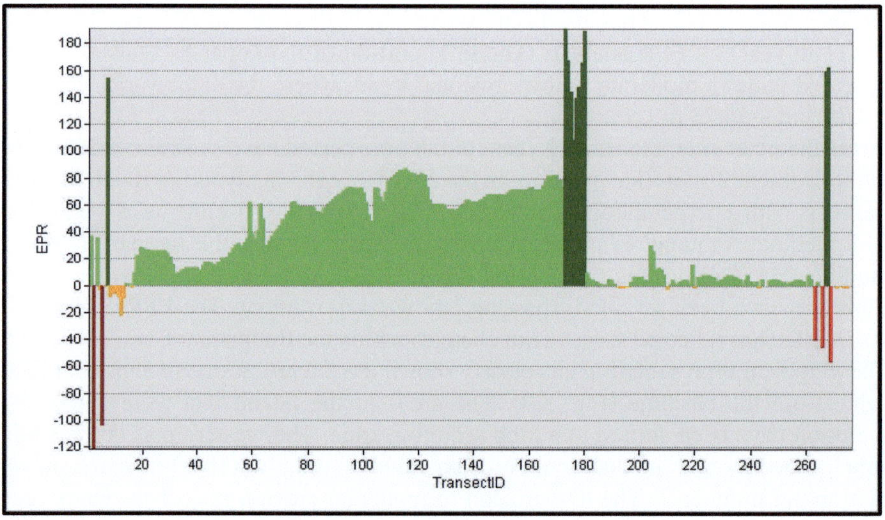

Fig. 3.7 Graph of Erosion-accretion rate by EPR (2000–2010), segment 2

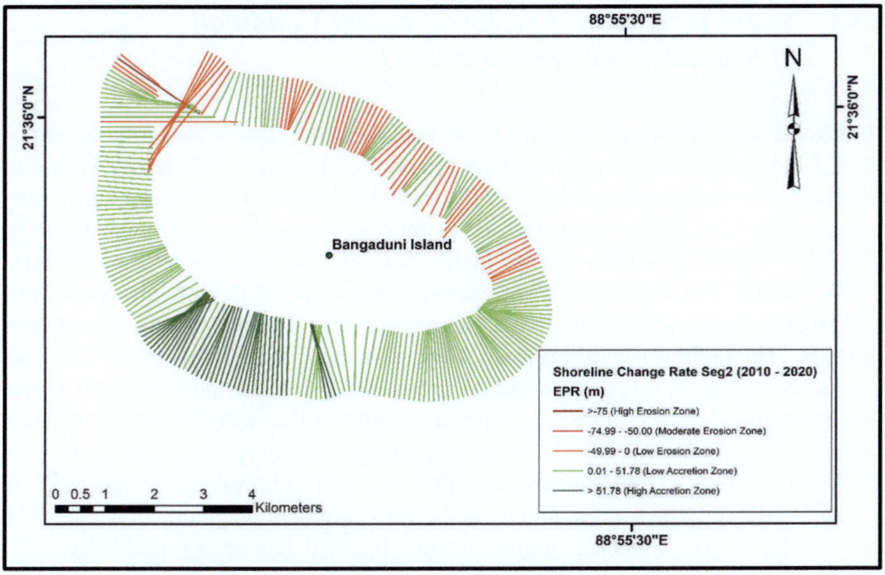

Fig. 3.8 Erosion-accretion zone of segment 2 (2010–2020)

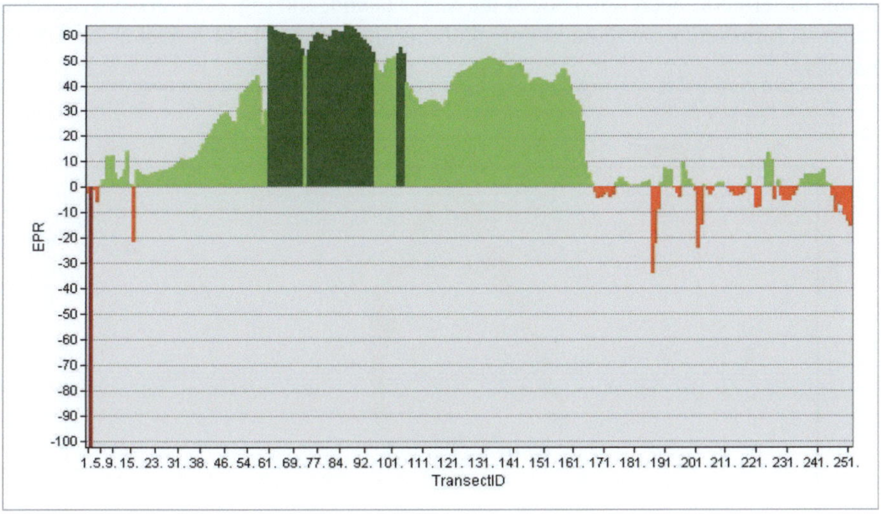

Fig. 3.9 Graph of Erosion-accretion rate by EPR (2010–2020), segment 2

3.3.3 Short-Term Shoreline Modification Condition of Segment 3 or Littoral Zone 3

The littoral zone 3 or segment 3 is the most important shoreline of Sundarbans, named Sagar Island, which is the most dynamic in nature. The first short-term changes of shoreline have been analyzed by the EPR model, a length of 75 km coastline using 715 transects with 2000 m perpendicular distance from baseline along a 750 m buffer area. This segment has recorded the major shoreline regression over the time span (2000—2010). The northern part of Sagar Island revealed the low, moderate, and high erosion zones, whereas the southern part was dominated by the low to moderate accretion. The highest negative (erosion) value has been reached by transect number 5, with the value of −120.44 m (12.04 m/year). During this period, Sagar Island has also been rich in accretion. The highest accretion value was obtained by transect number 399, with the value of 35.56 m. The high accretion trend has varied between 25.86 and 35.56 m, whereas the low accretion value varies between 2.93 and 23.76 m. On the other hand, the low erosion value by EPR has been obtained between −4.96 and −35.33 m. The moderate regression rate by EPR has been found between −40.75 and −77.87 m. The regression rate along the shoreline has been increasing in the mangrove ecosystem, and other studies also support this result, with net shoreline change rates in Sagar Island (Kumar et al., 2007; Sahana et al., 2019). Figure 3.10 reveals the erosion-accretion zone of the Sagar Island, and Fig. 3.11 also provides the graphical representation of shoreline alteration rate by the EPR model using DSAS analysis.

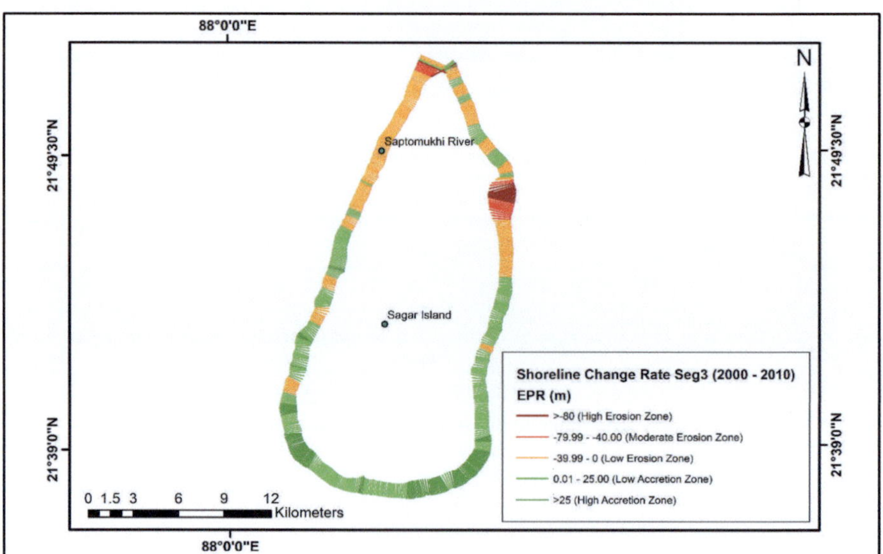

Fig. 3.10 Erosion-accretion zone of segment 3 (2000–2010)

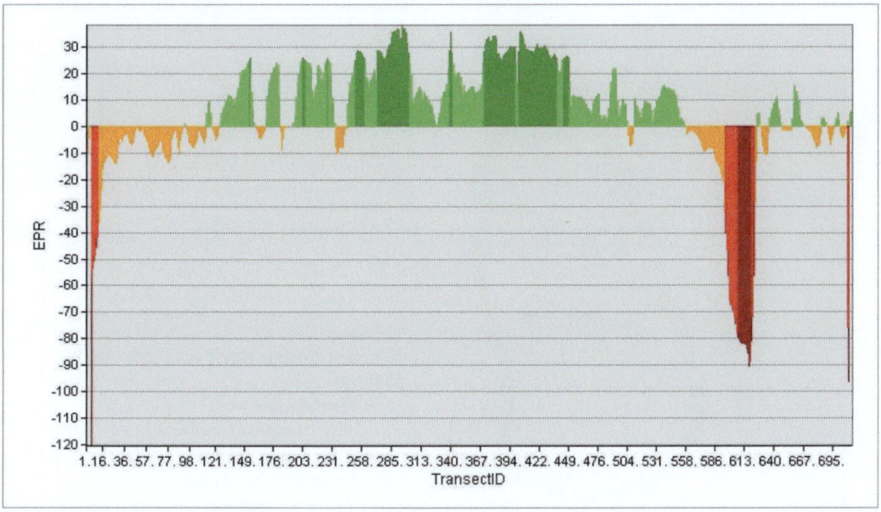

Fig. 3.11 Graph of Erosion-accretion rate by EPR (2000–2010), segment 3

On the other hand, the next short-term analysis has been performed between 2010 and 2020 time period using the DSAS analysis system through the EPR mode, where 719 transects were used to calculate the change rate. In this time period, it has been noticed that 30.89% of transects failed to represent the erosional trend with 3000 m perpendicular transects from baseline. In this case, it has been seen that areas that were previously within the low accretionary zone are now gradually being included in the low erosional zone. In 2000–2010, the western side of Sagar Island appeared to have more accretion, but in this case, it appears that the western part of the Island has started to have more erosional zone. On the other side, it has also been seen that the northern part, which was previously in a much more eroded zone, is now being included, and a large amount of the accretionary zone is found toward the north. The uppermost part of the west side of the Island has been exposed to a high regression trend (erosion), which is located near the Saptamukhi River estuary, with an average rate of −78.68 to −89. 45 m. The highest erosional rate has been observed as −109.6 m, with an annual rate of 10.96 m/year. However, another important thing to note here is that the high accretion zone has increased significantly compared to the previous short-term period (2000–2010). The area where high erosion was seen in the previous short-term period has now shifted to a large amount of high accretion, which is one of the important factors in mangrove growth. The low accretion rate has varied between 4.52 and 22.69 m. The low erosion rate has been observed between −2.79 and −24.58 m. The highest accretion rate has been found as 36.81 m. The erosion and accretion zone have been portrayed through Figs. 3.12 and 3.13, which reveal the erosion and accretion rate obtained by the EPR model.

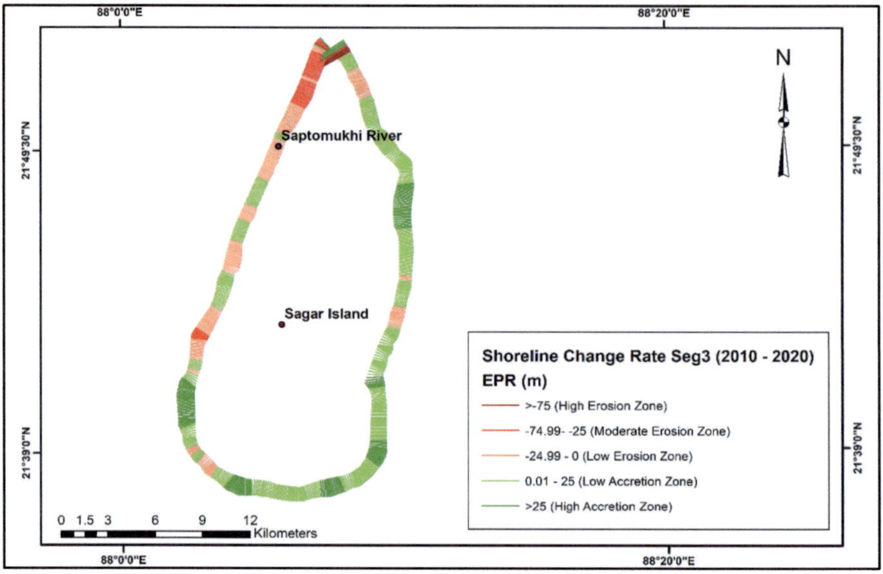

Fig. 3.12 Erosion-accretion zone of segment 3 (2010–2020)

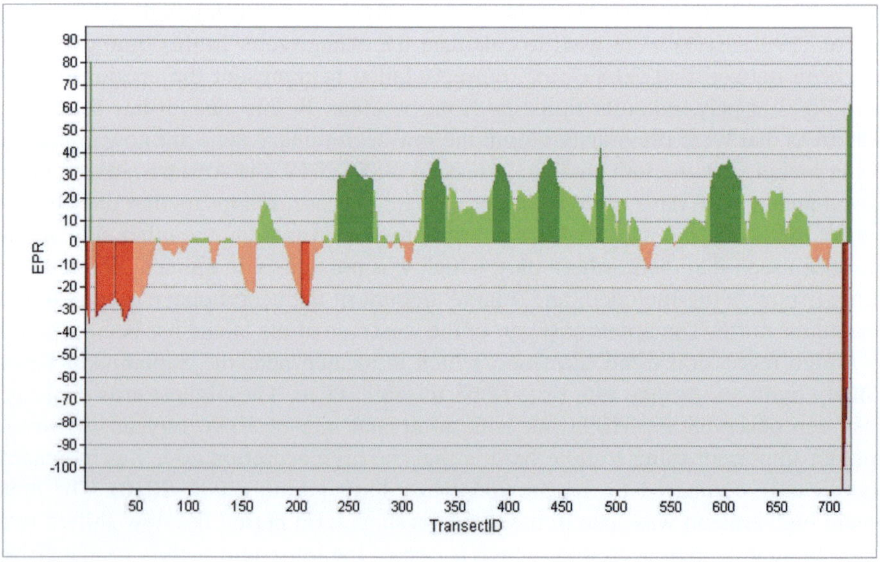

Fig. 3.13 Graph of Erosion-accretion rate by EPR (2010–2020), segment 3

3.3.4 Short-Term Shoreline Modification Condition of Segment 4 or Littoral Zone 4

One of the most significant southern islands is Murmu Island in the Indian Sundarbans, with a length of 25.07 km. The shoreline of this Island has been extracted, and the change rate has been calculated using the EPR model, where 271 transects were drawn to calculate the modification rate of the shoreline along with a 2000 m perpendicular line from the baseline. Here, 28 transects have been observed as erosional trends during the 2000–2010 time scale in the analysis. In the case of Murmu Island, the entire area appears to be included in the low accretion and some parts in the high accretionary regime. The high erosion zone has been observed on the northern part of the Murmu Island (Fig. 3.14). The maximum erosion rate was found in the upper part at 17.95 m by EPR. Figure 3.15 has revealed the erosion-accretion rate obtained by the EPR model. The high accretion rate has varied between 45.66 and 87.9 m in EPR. The highest accretion zone was noticed in the southern part of the Murmu Island.

On the other hand, 278 transects were generated to identify and estimate the erosion-accretion rate over Murmu Island between 2010 and 2020. In this case, it has been seen that in most parts of the Island, the low accretion zone shifted and turned into a low erosion zone (Fig. 3.16). In the northern part, where the estuary is connected, it has been observed that a large part has widened, and the amount of water inflow has also increased. A significant factor to note in the entire Island is that the high accretion zone is slowly decreasing and gradually shifting toward a low

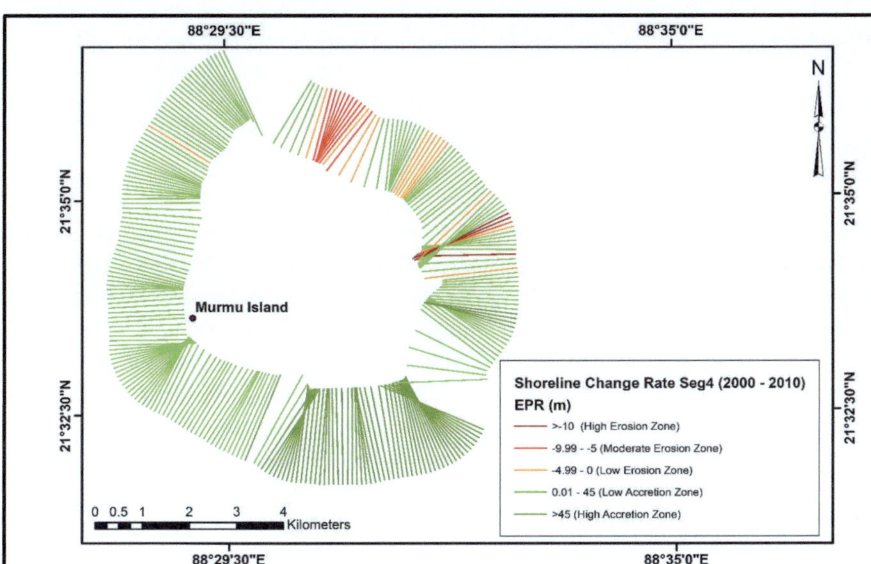

Fig. 3.14 Erosion-accretion zone of segment 4 (2000–2010)

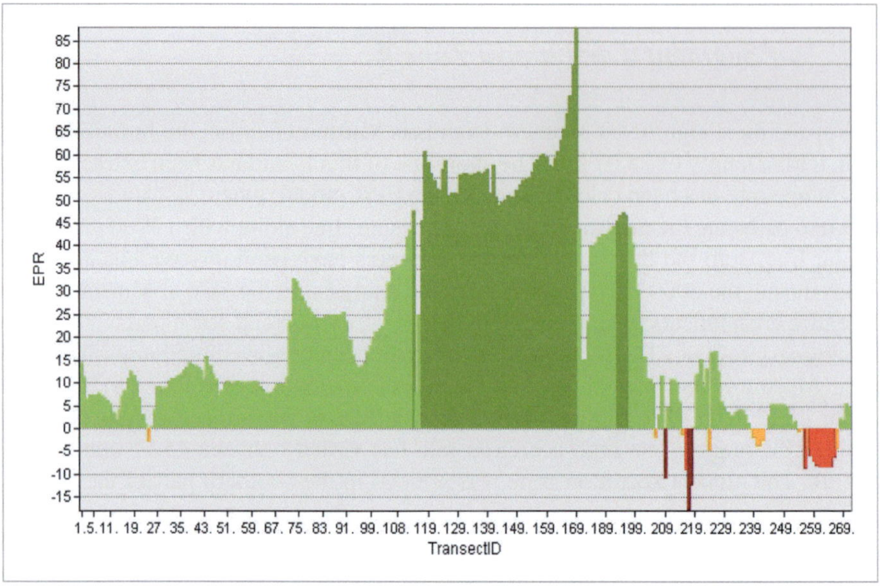

Fig. 3.15 Graph of Erosion-accretion rate by EPR (2000–2010), segment 4

accretion zone. This indicates that the amount of accretion is gradually decreasing and moving toward erosion. In this analysis, 50 transects have been used to indicate the erosion over the Island. The highest range of erosion was 55.05 m obtained in the EPR model. The average rate of low accretion has been found in the range of 7.95–45.85 m. Figure 3.17 reveals the erosion-accretion rate reached by EPR.

3.4 Discussion

Shoreline or coastline is a very important factor in terms of coastal sensitivity, which is influenced through the erosion and accretion process (Łabuz, 2015; Koroglu et al., 2019). The dynamicity of any delta shoreline is generally influenced by three main agents, i.e., sea level rise, sediment transportation, and tectonic and compaction subsidence (Tripathi et al., 2018). The shoreline position has directly modified the sediment transport and longshore wind effect (Rahman et al., 2011). The entire Indian Sundarbans has been directly controlled by the monsoon wind. Siltation is very common in the Hooghly River stretch, and recently, the Malta River has received a very poor sediment supply. The construction of the Farakka Dam has significantly reduced the supply of sediment in various parts of the Indian Sundarbans, which is one of the reasons for increasing erosion, and erosion is prominently visible on various islands (Wilson et al., 2017). In Bakkhali, a rampant erosion has been trapped due to the long shore sand transport near the estuary mouth, and due to this reason,

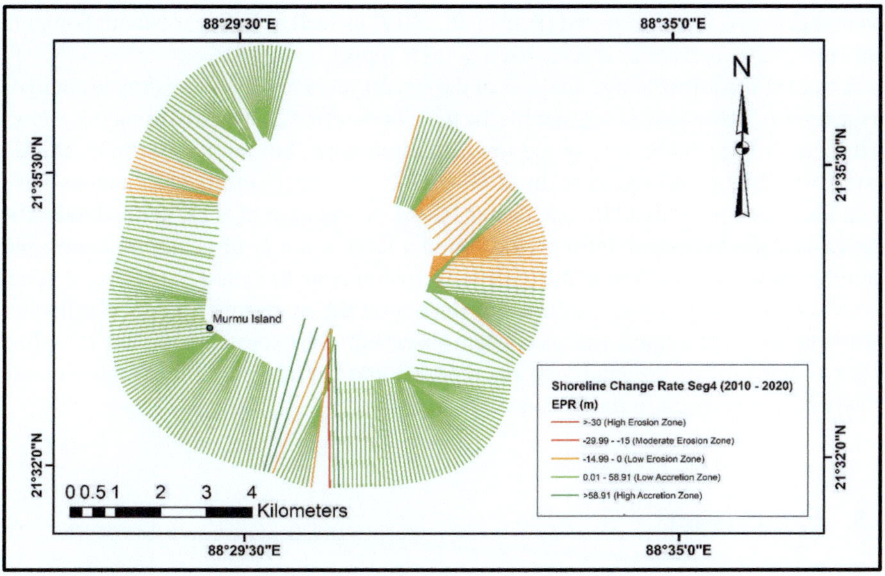

Fig. 3.16 Erosion-accretion zone of segment 4 (2010–2020)

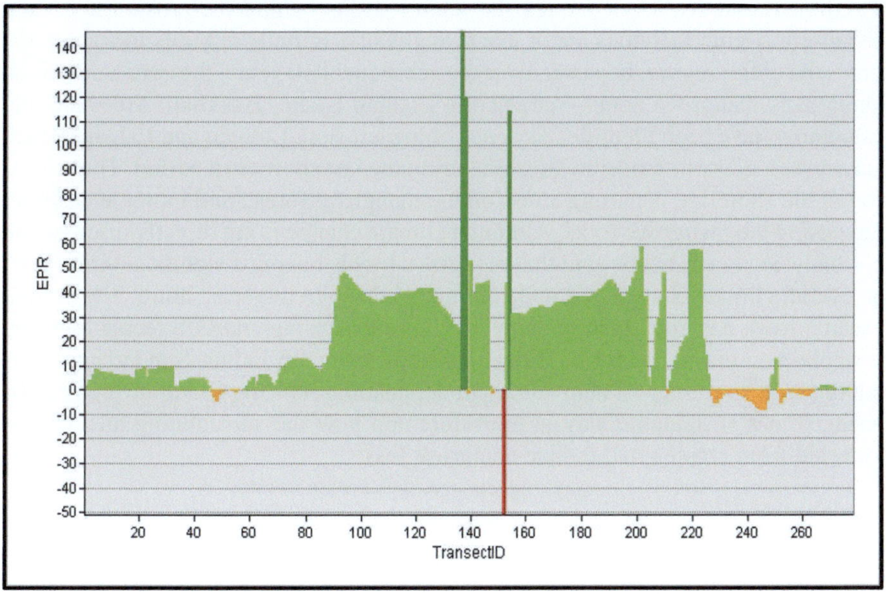

Fig. 3.17 Graph of Erosion-accretion rate by EPR (2010–2020), segment 4

the mangrove roots are exposed (Paul et al., 2017) as well as in Sagar Island. Samanta and Hazra (2017) have also revealed a similar report.

A brief short-term change analysis of the last 20 years reveals some drastic changes in the region. In the case of segment 1, the first short-term (2000–2010) analysis shows that there is a lot more area in the low accretion zone, but later it becomes shifted toward the low erosional zone in 2010–2020. The most significant changes have been identified in Bangaduni and Sagar Island. In the case of these two islands, the analysis of the first short-term period found a large number of accretion zones, but later the entire islands changed from a low erosion zone to a moderate erosion zone. Increased erosion has had a significant impact on the mangroves, and the saltwater intrusion over the region has also been observed. An excessive amount of saline water is harmful to the health of mangroves, and the roots of the mangroves are slowly being exposed to the outside.

3.5 Conclusion

The current result shows that the erosion and accretion are occurring in the same and simultaneous way. From the analysis, it has been seen that the low supply of sediment and sea level rise are the main reasons behind the erosion here, and the anthropogenic activities are at work. Segment 1 initially exhibits low accretion across the entire region; however, over the subsequent 10 years, the core areas of the Sundarbans mangrove zone—specifically Lothian Island, Bakkhali, Mousuni, and Kalasand—have been identified as zones of low erosion. Lothian and Lalganj Beach areas revealed low to moderate erosion during the last short-term period. The regression of the shoreline is putting stress on the mangrove forest, and increased salinity is exposing the mangrove roots. Although climate change is not directly discussed in this analysis, it can still be said that mangrove loss is having a significant impact on the overall climate change and will continue to do so in the near future. The current research work has attempted to show how remote sensing and GIS technology can be applied to the shoreline modification study over the Indian Sundarbans. This modification analysis will help a lot in understanding how the shoreline area can be managed in a sustainable way in the future and how the surrounding area can be protected from erosion and reduce mangrove loss.

References

Addo, K. A., & Kufogbe, K. S. (2011). Quantitative analysis of shoreline change using medium resolution satellite imagery in keta. *Ghana, 1*(1), 1–9.

Ahmed, W., & Neil, D. T. (1994). An evaluation of landsat thematic mapper digital data for discriminating coral reefzonation, heron reef (GPR). *International Journal of Remote Sensing, 15*(13), 2583–2597.

Alongi, D. M. (2015). The impact of climate change on mangrove forests. *Current Climate Change Reports, 1*, 30e39. https://doi.org/10.1007/s40641-015-0002-x.

Anand, R., Chandrasekar, B. N., Kaliraj, S., & Magesh, N. S. (2016). Shoreline change rate and erosion risk assessment along the Trou Aux Biches—Mont Choisy beach on the northwest coast of Mauritius using GIS-DSAS technique. *Environmental Earth Sciences, 75*(5), 1–12. https://doi.org/10.1007/s12665-016-5311-4.

Anbarasu, K., Baskaran, R., & Rajamanickam, G. V. (1999). Influence of sea level changes in the development of landforms around Chidambaram Tamilnadu. *Indian Journal Geomorph, 4*(1–2), 13–18.

Argyriou, A. V., Teeuw, R. M., Soupios, P., & Sarris, A. (2017). Neotectonic control on drainage systems: GIS-based geomorphometric and morphotectonic assessment for Crete, Greece. *Journal of Structural Geology, 104*, 93–111. https://doi.org/10.1016/j.jsg.2017.10.002.

Ayadi, K., Boutiba, M., Sabatier, F., & Guettouche, M. S. (2016). Detectionand analysis of historical variations in the shoreline, using digital aerialphotos, satellite images, and topographic surveys DGPS: Case of the Bejaia bay (East Algeria). *Arabian Journal of Geosciences, 9*(1), 26. https://doi.org/10.1007/s12517-015-2043-9.

Bera, R., & Maiti, R. (2019). Quantitative analysis of erosion and accretion (1975–2017) using DSAS-A study on Indian Sundarbans. *Regional Studies in Marine Science, 28*. https://www.researchgate.net/deref/ .

Bray, M. J., Carter, D. J., & Hooke, J. M. (1995). Littoral cell definition and budgets for central southern England. *Journal of Coastal Research*, 381–400. https://doi.org/10.1016/j.rsma.2023.103199.

Carter. V., & Schubert, J. (1974.). Coastal wetland analysis from ERTS-MSS digital data and field spectral measurement. In *Proceedings of the 9th International Symposium on Remote Sensing Environment, ERIM, Ann Arbor, MI, v2* (pp. 1279–1288).

Castelle, B., Masselink, G., Scott, T., Stokes, C., Konstantinou, A., Marieu, V., & Bujan, S. (2021). Satellite-derived shoreline detection at a high-energy meso-macrotidal beach. *Geomorphology, 383*, Article 107707. https://doi.org/10.1016/j.geomorph.2021.107707.

Cham, D. D., Son, T. N., Minh, Q. N., Thanh, T. N., & Dung, T. T. (2019). An analysis of shoreline changes using combined multitemporal remote sensing and digital evaluation model. *Civil Engineering Journal, 6*(1), 1–10.

Chatterjee, N., Mukhopadhyay, R., & Mitra, D. (2015). Decadal changes of shoreline pattern in Sundarbans, India. *Journal of Coastal Sciences, 2*(2), 54e64. http://drs.nio.org/drs/handle/2264/4847.

Chaudhuri, P., Chaudhuri, S., & Ghosh, R. (2019). The role of mangroves in coastal and estuarine sedimentary accretion in Southeast Asia. In: G. Aiello (Ed.), *Sedimentary processes-examples from Asia, Turkey and Nigeria*. Intech Open. https://doi.org/10.5772/intechopen.85591.

Cracknell, A. P., Mac, N., Fairlane, M. K., Chalton, J., & Ulbricht, K. A. (1982). Remote sensing in Scotland using data received from satellite—A study of the Tay estuary region using Landsat multispectral scanning imagery. *International Journal of Remote Sensing, 3*(2), 113–138.

Das, S., Choudhury, M. R., Das, S., & Khan, S. (2013). Monitoring shore line and Inland changes by using multi-temporal satellite data and risk assessment: A case study of Ghoramara Island. *International Journal of Geoscience and Technology, 1*(1), 1–20.

Das, S. K., Sajan, B., Ojha, C., & Soren, S. (2021). Shoreline change behavior study of Jambudwip island of Indian Sundarban using DSAS model. *The Egyptian Journal of Remote Sensing and Space Science, 24*(3), 961–970. https://doi.org/10.1016/j.ejrs.2021.09.004.

Deepika, B., Avinash, K., & Jayappa, K. S. (2014). Shoreline change rate estimation and its forecast: Remote sensing, geographical information system and statistics-based approach. *International Journal of Environmental Science and Technology, 11*(2), 395–416. https://doi.org/10.1007/s13762-013-0196-1.

Ding, Y. (2016). Sea-level rise and hazardous storms: Impact assessment on coasts and estuaries. In: W. Y. Chen, T. Suzuki, & M. Lackner (Eds.), Handbook of climate change mitigation and adaptation. New York: Springer. https://doi.org/10.1007/978-1-4614-6431-0_14-3.

Dolan, R., Fenster, M. S., & Holme, S. J. (1991). Temporal analysis of shoreline recession and accretion. *Journal of Coastal Research, 7*(3), 723–744.
Emran, A., Rob, M. A., Kabir, M. H., & Islam, M. (2016). Modeling spatio-temporal shoreline and areal dynamics of coastal island using geospatial technique. *Modeing Earth Systems and Environment, 2*(1), 4. https://doi.org/10.1007/s40808-015-0060-z.
Esmail, M., Mahmod, W. E., & Fath, H. (2019). Assessment and prediction of shoreline change using multi-temporal satellite images and statistics: Case study of Damietta coast, Egypt. *Applied Ocean Research, 82*, 274–282. https://doi.org/10.1016/j.apor.2018.11.009.
Fenster, M. S., Dolan, R., & Elder, J. F. (1993). A new method for predicting shoreline positions from historical data. *Journal of Coastal Research, 9*(1), 147–171.
Fletcher, C. H., Romine, B. M., Genz, A. S., Barbee, M. M., Dyer, M., Anderson, T. R., et al. (2012). National assessment of shoreline change: Historical shoreline change in the Hawaiian Islands: U.S. Geological Survey Open-File Report 2011–1051 (p. 55). https://pubs.usgs.gov/of/2011/1051/.
Gholami, M. D., & Baharlouii, M. (2019). Monitoring longterm mangrove shoreline changes along the northern coasts of the Persian Gulf and the Oman Sea. *Emerging Science Journal, 3*(2), 88. https://doi.org/10.28991/esj-2019-01172.
Gupta, L., Agrawal, N., Dixit, J., & Dutta, S. (2022). A GIS-based assessment of active tectonics from morphometric parameters and geomorphic indices of Assam Region, India. *Journal of Asian Earth Sciences: X, 8*, Article 100115. https://doi.org/10.1016/j.jaesx.2022.100115.
Hagenaars, G., de Vries, S., Luijendijk, A. P., de Boer, W. P., & Reniers, A. J. (2018). On the accuracy of automated shoreline detection derived from satellite imagery: A case study of the sand motor mega-scale nourishment. *Coastal Engineering, 133*, 113–125. https://doi.org/10.1016/j.coastaleng.2017.12.011.
Hale, R. P., Wilson, C. A., & Bomer, E. J. (2019). Seasonal variability of forces controlling sedimentation in the Sundarbans national forest, Bangladesh. *Frontiers in Earth Science, 7*, 211. https://doi.org/10.3389/feart.2019.00211.
Hill, G. J. E., Kelly, G. D., Tuiloma, P., & Phinn, S. R. (1994). Assessment of SPOT photographic imagery for mangrove in Figi. *Asian Pacific Remote Sensing Journal, 7*(1), 1–7.
Himmelstoss, E. A., Henderson, R. E., Kratzmann, M. G., & Farris, A. S. (2021). Digital shoreline analysis system (DSAS) version 5.1 user guide (No. 2021–1091). US Geological Survey.
Hossain, M. S., Yasir, M., Wang, P., Ullah, S., Jahan, M., Hui, S., & Zhao, Z. (2021). Automatic shoreline extraction and change detection: A study on the southeast coast of Bangladesh. *Marine Geology, 441*, Article 106628. https://doi.org/10.1016/j.margeo.2021.106628.
Hwang, C. S., Choi, C. U., & Choi, J. S. (2014). Shoreline changes interpreted from multi-temporal aerial photographs and high-resolution satellite images. A case study in Jinha beach. *Korean Journal of Remote Sensing, 30*(5), 607–616. https://doi.org/10.7780/kjrs.2014.30.5.6.
Islam, A. M., & Hasan, H. (2020). Generation of IDF equation from catchment delineation using GIS. *Civil Engineering Journal, 6*(3), 540–547. https://doi.org/10.28991/cej-2020-03091490.
Jana, A., Biswas, A., Maiti, S., & Bhattacharya, A. (2013). Shoreline changes in response to sea level rise along Digha Coast, Eastern India: An analytical approach of remote sensing, GIS and statistical techniques. *Journal of Coastal Conservation, 18*(3), 145–155. https://doi.org/10.1007/s11852-013-0297-5.
Jana, A., Maiti, S., & Biswas, A. (2016). Analysis of short-term shoreline oscillations along Midnapur-Balasore Coast, Bay of Bengal, India: A study based on geospatial technology. *Modeling Earth Systems and Environment, 2*(2), 64. https://doi.org/10.1007/s40808-016-0117-7.
Jayappa, K. S., Mitra, D., & Mishra, A. K. (2006). Coastal geomorphological and land-use and land-cover study of Sagar Island, Bay of Bengal (India) using remotely sensed data. *International Journal of Remote Sensing, 27*(17), 3671–3682. https://doi.org/10.1080/01431160500500375.
Kabuth, A. K., Kroon, A., & Pedersen, J. B. T. (2013). Multidecadal shoreline changes in Denmark multidecadal shoreline changes in Denmark. *Journal of Coastal Research, 30*(4), 714–728.

Kaliraj, S., Chandrasekar, N., & Magesh, N. S. (2015). Evaluation of coastal erosion and accretion processes along the southwest coast of Kanyakumari, Tamil Nadu using geospatial techniques. *Arabian Journal of Geosciences, 8*(1), 239–253. https://doi.org/10.1007/s12517-013-1216-7.

Karunarathna, H., Brown, J., Chatzirodou, A., Dissanayake, P., & Wisse, P. (2018). Multitimescale morphological modelling of a dune-fronted sandy beach. *Coastal Engineering, 136*, 161e171. https://doi.org/10.1016/j.coastaleng.2018.03.005.

Kassakian, J., Jones, A., Martinich, J., & Hudgens, D. (2017). Managing for no net loss of ecological services: An approach for quantifying loss of coastal wetlands due to sea level rise. *Environmental Management, 59*(5), 736e751. https://doi.org/10.1007/s00267-016-0813-0.

Kelley, G. W., Hobgood, J. S., Bedford, K. W., & Schwab, D. J. (1998). Generation of three-dimensional lake model forecasts for Lake model forecasts for Lake Erie. *Journal Weather and Forecasting, 13*(3), 305–315.

Kermani, S., Boutiba, M., Guendouz, M., Guettouche, M. S., & Khelfani, D. (2016). Detection and analysis of shoreline changes using geospatial tools and automatic computation: Case of Jijelian sandy coast (East Algeria). *Ocean & Coastal Management, 132*, 46–58. https://doi.org/10.1016/j.ocecoaman.2016.08.010.

Kona, A. D., Hoque, M. E., & Rahman, M. A. (2025). Evaluating shoreline prediction accuracy with the Kalman filter model: A case study of Nijhum Dwip, Bay of Bengal. *Remote Sensing Applications: Society and Environment, 37*, Article 101469. https://doi.org/10.1016/j.rsase.2025.101469.

Konstantinou, A., Scott, T., Masselink, G., Stokes, K., Conley, D., & Castelle, B. (2023). Satellite-based shoreline detection along high-energy macrotidal coasts and influence of beach state. *Marine Geology, 462*, Article 107082. https://doi.org/10.1016/j.margeo.2023.107082.

Koroglu, A., Ranasinghe, R., Jiménez, J. A., & Dastgheib, A. (2019). Comparison of coastal vulnerability index applications for Barcelona Province. *Ocean & Coastal Management, 178*, Article 104799. https://doi.org/10.1016/j.ocecoaman.2019.05.001.

Kulp, S. A., & Strauss, B. H. (2019). New elevation data triple estimates of global vulnerability to sea-level rise and coastal flooding. *Nature Communications, 10*, 4844. https://doi.org/10.1038/s41467-019-12808-z.

Kumar, P. D., Gopinath, G., Laluraj, C. M., Seralathan, P., & Mitra, D. (2007). Change detection studies of Sagar Island, India, using Indian remote sensing satellite 1C linear imaging self-scan sensor III data. *Journal of Coastal Research, 23*(6), 1498–1502. https://doi.org/10.2112/05-0599.1.

Kumar, S., Imen, S., Sridharan, V. K., Gupta, A., McDonald, W., Ramirez-Avila, J. J., & Baskaran, L. (2024). Perceived barriers and advances in integrating earth observations with water resources modeling. *Remote Sensing Applications: Society and Environment, 33*, Article 101119. https://doi.org/10.1016/j.rsase.2023.101119.

Łabuz, T. A. (2015). Environmental impacts—Coastal erosion and coastline changes. In: The BACC II Author Team (Eds.), *Second assessment of climate change for the Baltic Sea Basin. Regional climate studies*. Cham: Springer. https://doi.org/10.1007/978-3-319-16006-1_20.

Maiti, S., & Bhattacharya, A. K. (2009). Shoreline change analysis and its application to prediction: A remote sensing and statistics-based approach. *Marine Geology, 257*(1–4), 11–23. https://doi.org/10.1016/j.margeo.2008.10.006.

Mansour, N., Sarhan, T., El-Gamal, M., & Nassar, K. (2024). Assessing the compatibility of EPR rates with one-dimensional numerical modelling in monitoring shoreline kinematics along with supplying near/long-term forecasts. *Regional Studies in Marine Science, 71*, Article 103391. https://doi.org/10.1016/j.rsma.2024.103391.

Marghany, M., Sabu, Z., & Hashim, M. (2010). Mapping coastal geomorphology changes using synthetic aperture radar data. *International Journal of Physical Sciences, 5*(12), 1890–1896.

Masria, A., Nadaoka, K., Negm, A., & Iskander, M. (2015). Detection of shoreline and land cover changes around Rosetta promontory, Egypt, based on remote sensing analysis. *Land, 4*(1), 216–230. https://doi.org/10.3390/land4010216.

Maurya, D. M., Thakkar, M. G., Patidar, A. K., Bhandari, S., Goyal, B., & Chamyal, L. S. (2010). Late quaternary geomorphic evolution of the coastal zone of Kachchh, Western India. *Journal of Coastal Research, 24*(3), 746–758.

Maurya, D. M., Tiwari, P., Shaikh, M., Patidar, A. K., Vanik, N., Padmalal, A., & Chamyal, L. S. (2021). Late quaternary drainage reorganization assisted by surface faulting: The example of the Katrol Hill Fault zone, Kachchh, Western India. *Earth Surface Processes and Landforms, 46*(7), 1268–1293. https://doi.org/10.1002/esp.5097.

Mishra, M., Chand, P., Pattnaik, N., Kattel, D. B., Panda, G. K., Mohanti, M., et al. (2019). Response of long-to short-term changes of the Puri coastline of Odisha (India) to natural and anthropogenic factors: A remote sensing and statistical assessment. *Environmental Earth Sciences, 78*(11), 338. https://doi.org/10.1007/s12665-019-8336-7.

Mishra, M., Kar, P. K., Chand, P., Mohanty, P. K., Acharyya, T., Santos, C. A. G., & Gonçalves, R. M. (2023). Deciphering the impact of anthropogenic coastal infrastructure on shoreline dynamicity along Gopalpur coast of Odisha (India): An integrated assessment with geospatial and field-based approaches. *Science of the Total Environment, 858*, Article 159625. https://doi.org/10.1016/j.scitotenv.2022.159625.

Mujabar, P. S., & Chandrasekar, N. (2011). Shoreline change analysis along the coast between Kanyakumari and Tuticorin of India using remote sensing and GIS. *Arabian Journal of Geosciences*. https://doi.org/10.1007/s12517-011-0394-4.

Mukhopadhyay, A., Mukherjee, S., Mukherjee, S., Ghosh, S., Hazra, S., & Mitra, D. (2012). Automatic shoreline detection and future prediction: A case study on Puri Coast, Bay of Bengal, India. *European Journal of Remote Sensing, 45*(1), 201–213.

Murali, R. M., Dhiman, R., Choudhary, R., Seelam, J. K., Ilangovan, D., & Vethamony, P. (2015). Decadal shoreline assessment using remote sensing along the central Odisha coast, India. *Environmental Earth Sciences, 74*(10), 7201–7213. https://doi.org/10.1007/s12665-015-4698-7.

Muskananfola, M. R., & Febrianto, S. (2020). Spatio-temporal analysis of shoreline change along the coast of Sayung Demak, Indonesia using Digital Shoreline Analysis System. *Regional Studies in Marine Science, 34*, Article 101060. https://doi.org/10.1016/j.rsma.2020.101060.

Nandi, S., Ghosh, M., Kundu, A., Datta, D., & Baksi, M. (2016). Shoreline shifting and its prediction using remote sensing and GIS techniques: A case study of Sagar Island, West Bengal (India). *Journal of Coastal Conservation, 20*, 61–80. https://doi.org/10.1007/s11852-015-0418-4

Nassar, K., Mahmod, E. W., Fath, H., Masria, A., Nadaoka, K., & Negm, A. (2018). Shoreline change detection using DSAS technique: Case of North Sinai Egypt. *Marine Geo Resources & Geotechnology*. https://doi.org/10.1080/1064119X.2018.1448912.

Nath, A., Koley, B., Saraswati, S., & Ray, B. C. (2021). Identification of the coastal hazard zone between the areas of Rasulpur and Subarnarekha estuary, east coast of India using multi-criteria evaluation method. *Modeling Earth Systems and Environment, 7*(4), 2251–2265. https://doi.org/10.1007/s40808-020-00986-5.

Nath, A., Koley, B., Saraswati, S., Choudhury, T., Um, J.-S., & Ray, B. C. (2023). Geospatial analysis of short-term shoreline change behavior between Subarnarekha and Rasulpur estuary, east coast of India using intelligent techniques (DSAS). *GeoJournal, 88*(Suppl 1), 255–275. https://doi.org/10.1007/s10708-022-10683-8.

Nath, A., Koley, B., Saraswati, S., Choudhury, T., Um, J.-S., & Ray, B. C. (2024a). Geo-spatial analysis of erosion accretion dynamics and areal changes along Digha Coastline in West Bengal, India. *Remote Sensing in Earth Systems Sciences, 7*, 183–205. https://doi.org/10.1007/s41976-024-00114-2.

Nath, A., Koley, B., Choudhury, T., Saraswati, S., & Um, J.-S. (2024b). Coastal vulnerability differentiated according to geospatial quartiles method: Rasulpur to Subarnarekha estuary, east coast of India. *Spatial Information Research, 32*, 383–398. https://doi.org/10.1007/s41324-023-00554-w.

Narayana, A. C. (2016). Shoreline changes. In *Encyclopedia of estuaries*. Springer, Dordrecht. 590–602.

Nayak, S. R. (2002). Use of satellite data in coastal mapping. *Indian Cartographer, 22*, 147–157.

Nayak, S. R., & Sahai, B. (1985). Coastal morphology: A case study of the gulf of Khambhat (Cambay). *International Journal of Remote Sensing, 6*(3–4), 559–567.

Passeri, D. L., Hagen, S. C., Medeiros, S. C., Bilskie, M. V., Alizad, K., & Wang, D. (2015). The dynamic effects of sea level rise on low-gradient coastal landscapes: A review. *Earth's Future, 3*, 159e181. https://doi.org/10.1002/2015EF000298.

Paul, S., Mishra, M., Guria, R., Pati, S., Baraj, B., Marques da Silva, R., & Guimaraes Santos, C. (2024). A multi-temporal analysis of shoreline dynamics influenced by natural and anthropogenic factors: Erosion and accretion along the Digha Coast, West Bengal, India. *Marine Pollution Bulletin, 200*, Article 116089. https://doi.org/10.1016/j.marpolbul.2024.116089.

Paul, A., Ray, R., Kamila, A., & Jana, S. (2017). Mangrove degradation in the Sundarbans. Coastal wetlands: Alteration and remediation. *Coastal Research Library, 21*. https://doi.org/10.1007/978-3-319-56179-0_11.

Pranzini, E., Cinelli, I., Cipriani, L. E., & Anfuso, G. (2020). An integrated coastal sediment management plan: The example of the Tuscany region (Italy). *Journal of Marine Science and Engineering, 8*, 33. https://doi.org/10.3390/jmse8010033.

Rahman, F., Dragoni, D., & El Masri, B. (2011). Response of the Sundarbans coastline to sea level rise and decreased sediment flow: A remote sensing assessment. *Remote Sensing of Environment, 115*, 3121. https://doi.org/10.1016/j.rse.2011.06.019.

Ramachandran, A., Sujatha, M., Alruwais, N., & Alshahrani, H. M. (2025). Forecasting coastal stability: Digital shoreline analysis system and machine learning techniques in evaluating impacts of cyclones. *Regional Studies in Marine Science, 81*, Article 103961. https://doi.org/10.1016/j.rsma.2024.103961.

Ramalingam, G., & Renganathan, M. (1998). Coastal geomorphology and its evolution—A case study in Agniar basin Tamilnadu. *Journal of Indian Geomorphology, 3*(2), 146–156.

Ramon, C. M., Capa, M. S., Saavedra, H. F., & Paredes, J. R. (2019). Integrated coastal zone management in continental Ecuador and Galapagos islands: Challenges and opportunities in a changing tourism and economic context. *Sustainability, 11*(22), 6386. https://doi.org/10.3390/su11226386.

Romine, B. M., & Fletcher, C. H. (2013). A summary of historical shoreline changes on beaches of Kauai, Oahu, and Maui, Hawaii. *Journal of Coastal Research, 29*(3), 605–614.

Ryu, J. H., Won, J. S., & Min, K. D. (2002). Waterline extraction from Landsat TM data in a tidal flat: A case study in Gosmo Bay. *Korea. Remote Sensing of Environment, 83*(3), 442–456. https://doi.org/10.1016/S0034-4257(02)00059-7.

Sahana, M., Hong, H., Ahmed, R., Patel, P., Bhakat, P., & Sajjad, H. (2019). Assessing coastal Island vulnerability in the Sundarban Biosphere Reserve, India, using geospatial technology. *Environmental Earth Sciences, 78* https://doi.org/10.1007/s12665-019-8293-1.

Samanta, K., & Hazra, S. (2017). Mangrove forest cover changes in Indian Sundarban (1986–2012) using remote sensing and GIS. In S. Hazra, A. Mukhopadhyay, A. Ghosh, D. Mitra, & V. Dadhwal (Eds.), *Environment and earth observation. Springer remote sensing/photogrammetry*. Cham: Springer. https://doi.org/10.1007/978-3-319-46010-9_7.

Sánchez-García, E., Palomar-Vázquez, J. M., Pardo-Pascual, J. E., Almonacid-Caballer, J., Cabezas-Rabadán, C., & Gómez-Pujol, L. (2020). An efficient protocol for accurate and massive shoreline definition from mid-resolution satellite imagery. *Coastal Engineering, 160*, Article 103732. https://doi.org/10.1016/j.coastaleng.2020.103732.

Saranathan, E., Chandrasekaran, R., Soosai Manickaraj, D., & Kannan, M. (2011). Shoreline changes in Tharangampadi village, Nagapattinam district, Tamil Nadu, India—A case study. *Journal of the Indian Society of Remote Sensing, 39*(1), 107–115. https://doi.org/10.1007/s12524-010-0052-4.

Siddiqui, M. N., & Maajid, S. (2004). Monitoring of geomorphological changes for planning reclamation work in coastal area of Karachi, Pakistan. *Advances in Space Research, 33*(7), 1200–1205. https://doi.org/10.1016/S0273-1177(03)00373-9.

Singh, A. (1989). Digital change detection techniques using remotely sensed data. *International Journal of Remote Sensing, 10*(6), 989–1003. https://doi.org/10.1080/01431168908903939.

Suresh, P. K., & Sundar, V. (2011). Comparison between measured and simulated shoreline changes near the tip of Indian Peninsula. *Journal of Hydro-Environment Research, 5*(3), 157–167. https://doi.org/10.1016/j.jher.2011.02.002.

Taib, H., Hadji, R., & Hamed, Y. (2024). Exploring neotectonic activity in a semiarid basin: A case study of the Ain Zerga watershed. *Journal of Umm Al-Qura University for Applied Sciences, 10*, 20–33. https://doi.org/10.1007/s43994-023-00072-3.

Thakur, S., Dey, D., Das, P., Ghosh, P. B., & De, T. K. (2017). Shoreline change detection using remote sensing in the Bakkhali coastal region, West Bengal, India. *Indian Journal of Geosciences, 71*(4), 611–626.

Thakur, S., Mondal, I., Bar, S., Nandi, S., Ghosh, P. B., Das, P., & De, T. K. (2021). Shoreline changes and its impact on the mangrove ecosystems of some islands of Indian Sundarbans, North-East coast of India. *Journal of Cleaner Production, 284*, Article 124764. https://doi.org/10.1016/j.jclepro.2020.124764.

Thakur, S., Mondal, I., Das, P., Ghosh, P. B., & De, T. K. (2018). Time series analysis of shoreline oscillations and its repercussions on the mangrove's ecosystems around Bakkhali island of Sundarban delta, Eastern India. In *International Conference Biodiverse 2018 at: IIT Guwahati Affiliation* (Vol. 71, p. 611e626). Indian Institute of Technology Guwahati. https://doi.org/10.13140/RG.2.2.33507.66088/1.

Thieler, E. R., Himmelstoss, E. A, Zichichi, J. L., & Miller, T. L. (2005). *The Digital Shoreline Analysis System (DSAS), an ArcGIS extension for calculating historic shoreline change*. https://doi.org/10.3133/ofr20051304.

Thieler, E. R., Himmelstoss, E. A., Zichichi, J. L., & Ergul, A. (2019). *The Digital Shoreline Analysis System (DSAS) version 4.0 an ArcGIS extension for calculating shoreline change*. US Geological Survey.

Tran Thi, V., Tien Thi Xuan, A., Phan Nguyen, H., Dahdouh-Guebas, F., & Koedam, N. (2014). Application of remote sensing and GIS for detection of long-term mangrove shoreline changes in Mui Ca Mau Vietnam. *Biogeosciences, 11*(14), 3781–3795. https://doi.org/10.5194/bg-11-3781-2014.

Tripathi, S., Resmi, S., & Baraik, S. (2018). Substantial land subsidence and its impact on Kalinagar, Sundarban delta, West Bengal. *Indian Journal of Geosciences, 71*(4), 589–598.

Tsai, Y. L. S. (2024). Nation-scale multidecadal shoreline extraction and coastal spatio-temporal change monitoring using cross-mission remote sensing data. *Ocean & Coastal Management, 253*, Article 107136. https://doi.org/10.1016/j.ocecoaman.2024.107136.

Valderrama-Landeros, L., Flores-de-Santiago, F., Álvarez-Sánchez, L. F., Flores-Verdugo, F., & Rodríguez-Sobreyra, R. (2024). The influence of spatial resolution on coastline detection by means of multisource remote sensing data. *Remote Sensing Applications: Society and Environment, 35*, Article 101258. https://doi.org/10.1016/j.rsase.2024.101258.

Vaz, E., & Bowman, L. (2013). An application for regional coastal erosion processes in Urban Areas; A case study of the golden Horseshoe in Canada. *Land, 2*, 595e608. https://doi.org/10.3390/land2040595.

Wagner, T. W., Michalek, J. L., & Laurin, R. (1991). Remote sensing application in the coastal zone: A case from the Dominican Republic. In R. W. Stoffle & D. D. Halmo (Eds.), *International earth science information network report* (pp. 17–56). University Center.

Wang, X. Z., Zhang, H. G., Fu, B., & Shi, A. (2013). An alysison the coastline change and erosion-accretion evolution of the Pearl River estuary, China, based on remote-sensing images and nautical charts. *Journal of Applied Remote Sensing, 7*(1), Article 073519. https://doi.org/10.1117/1.JRS.7.073519.

Ward, R. D., Friess, D. A., Day, R. H., & MacKenzie, R. A. (2016). Impacts of climate change on mangrove ecosystems: A region by region overview. *Ecosystem Health and sustainability, 2*(4), Article e01211. https://doi.org/10.1002/ehs2.1211.

Williams, A., Rangel-Buitrago, N., Pranzini, E., & Anfuso, G. (2018). The management of coastal erosion. *Ocean and Coastal Management, 156*, 4. https://doi.org/10.1016/j.ocecoaman.2017.03.022.

Wilson, C., Goodbred, S., Small, C., Gilligan, J., Sams, S., Mallick, B., et al. (2017). Widespread infilling of tidal channels and navigable waterways in humanmodified tidal deltaplain of Southwest Bangladesh. *Elementa: Science of the Anthropocene, 5*, 78. https://doi.org/10.1525/elementa.263.

Woodworth, P. L., Melet, A., Marcos, M., Ray, R. D., Wopplemann, G., Sasaki, Y. N., et al. (2019). Forcing factors affecting sea level changes at the coast. *Surveys in Geophysics, 40*, 1351e1397. https://doi.org/10.1007/s10712-019-09531-1.

Yamano, H., Shimazaki, H., Matsunaga, T., Ishoda, A., McClennen, C., Yokoki, H., et al. (2006). Evaluation of various satellite sensors for waterline extraction in a coral reef environment: Majuro Atoll, Marshall Islands. *Geomorphology, 82*(3–4), 398–411. https://doi.org/10.1016/j.geomorph.2006.06.003.

Yanli, T. (2011). *The application of GIS and RS for coastline change detection and risk assessment to enhanced sea level rise.* International Institute for Geo information Science and Earth Observation.

Chapter 4
Evaluating Long-Term Shoreline Shifts in the Indian Sundarbans Using Digital Shoreline Analysis System (DSAS)

Abstract Coastline dynamics reflect the complex interrelation between the coastal processes, sea level variations, and human activities, making their long-term monitoring essential for sustainable coastal zone management. The current chapter uses the Digital Shoreline Analysis System (DSAS) through a GIS application, applying the multi-decadal satellite data to obtain the long-term shoreline modification condition over the Indian Sundarbans part along various islands. Multi-resolution satellite data have been used in this analysis to obtain the change rate using the statistical method of Linear Regression Rate (LRR), where more than two shorelines are calculated to reaching out the change rate. The present study applies the method during the 2000–2020 time period. The study highlights the long-term regression and transgression of the shoreline with a specific change rate. Long-term change of shoreline reveals the cumulative impacts of physical forces such as waves, storm surges, tides, sea level rise, and sediment transport, along with anthropogenic pressures including coastal infrastructure development and land reclamation. The results indicate a scientific basis for hazard zonation, policy-making formulation, and the possibility of adaptive mitigation strategies aimed at emphasizing coastal resilience in the context of climate change and rising sea levels.

Keywords Long-term Analysis · Digital Shoreline Analysis System (DSAS) · Linear Regression Rate (LRR) · Shoreline Modification · Littoral Zones · Geospatial Techniques · Indian Sundarbans

4.1 Introduction

The dynamicity nature of the coastal region and the modification of the shoreline have been widely creating an immense effect on various factors of the environment, already elaborated by several authors (Chowdhury et al., 2023; Pranzini & Williams, 2013; Saddam et al., 2025; Thomas et al., 2010; Woodroffe, 2002). The wind direction and wave height have also been included in these factors of the environment (Thomas et al., 2011; Wang et al., 2022). Few studies on shorelines have also shown

© The Author(s), under exclusive license to Springer Nature Switzerland AG 2026

A. Nath et al., *Shoreline Modification Over the Indian Sundarban*,
SpringerBriefs in GIS, https://doi.org/10.1007/978-3-032-13981-8_4

that some variables are very transient and modifications to the morphology may be changed after eased of any triggering factors in the coastline (Mahapatra et al., 2014a, b; Wright & Thom, 2023). Coast occupied areas are always changing in nature. These changes are required to be monitored and redefined constantly. The changes in coastal areas occurred in several modifications, i.e., coastal soil drifting in the sea, river deltas directly affecting the shoreline, and environmental changes resulting from geographical and morphological conditions, as well as sea level rise (Apostolopoulos & Nikolakopoulos, 2021; Laignel et al., 2023; Nidhinarangkoon et al., 2023; Parthasarathy & Deka, 2019; Rahman et al., 2022). Expansion of settlements near the shoreline has been triggering the erosion that leads to land regression, and it tends to modify the coastal region over time period (Burningham & French, 2017; Dev et al., 2023).

In case of sustainable management for decision-making policies in any coastal region, it is most imperative to consider the natural and socio-economic impacts on the coastal region (Dasgupta et al., 2009; Hossain et al., 2021; Lakshmi & Rajagopalan, 2000; Yasir et al., 2020, 2021). To predict the past, present, and future condition of the coastline, coastal engineering and management are very crucial. The shoreline modification data is very much imperative in several aspects, i.e., validating the coastal numerical models (Hanson et al., 1988); shore protection engineering (Apostolopoulos et al., 2022), sea level rise assessment (Leatherman, 2001), etc. It is essential to define the coastal boundaries in both conditions of time and space due to the dynamic nature of shoreline boundaries. After various discussions, it has been concluded that a prominent definition is that the interface zone of land and water is considered as shoreline or coastline (Tanwari et al., 2025; Strayer & Findlay, 2010; Whitfield & Elliott, 2011), but this definition is not obtained promptly; it is actually challenging. The coastline position is continuously changed or modified over a time period due to the sediment transport, constant water level movement, wave and tide actions, storm surges, and other factors. For the evaluation of spatial and temporal changes in shoreline, the dynamic nature of boundaries is very imperative (Zheng et al., 2023). The prompt location of coastline refers to the immediate spatial condition of the land–water interface (Hui et al., 2022; Mullick et al., 2020). It is imperative to acknowledge that the shore areas are time-representing factors with long- and short-term scales. Also, it is important to consider the coastal morphology when accounting for the coastal changes.

Coastal areas are most significant where the hydrosphere, atmosphere, and earth factors are directly associated, and during coastal monitoring, all these factors need to be considered, comparing with recent data to historical data for coastal zone management using remote sensing and Geographical Information System (GIS) technologies. Remote sensing data has been considered the most applicable data source while performing any coastal change analysis for management and planning studies (Yasir et al., 2024; Uckac, 1998). A prominent idea about coastal location is required while assessing the shoreline change or modification and identification of land-sea interface conditions and related resources. But this is a very challenging, time-consuming method with conventional technologies, as ground truth data analysis with a vast coastal region (Kvile et al., 2024; Ciampalini et al., 2015; Maurya et al., 2021). Significant changes in the shoreline have been promptly delivered by remote sensing data (Ghaderi & Rahbani, 2020; Hossen & Sultana, 2023; Sekar et al., 2024).

One of the most applicable and versatile change detection tools has been considered as DSAS (Digital Shoreline Analysis System), which is widely used for coastal monitoring systems (Baig et al., 2020; Hossen & Sultana, 2023; Zorlu & Kusak, 2025). In coastal areas, long-term analysis has the potential to identify modifications to the coastline using several methods. Various methods are included in coastal change analysis, i.e., historic or past data analysis, aerial image analysis, and numerous topographical map evaluation techniques (Burningham & French, 2017; Esteves et al., 2002; Granja & Carvalho, 2000; Kaliraj et al., 2017; Silva & Duck, 2006; Thomas et al., 2010). As many authors have pointed out, coastline marking is a fundamental aspect to consider in mapping coastline diversity and identifying the perfect location of shoreline, not just in images or on the earth's surface (Douglas & Crowell, 2000; Parker, 2003; Thomas et al., 2010, 2011; Zhang et al., 2002). The high waterline has been defined by the actual shoreline position. But the vegetation line is mostly applied in both aerial and historical images while collecting the coastline because it is easily perceived (Rogers et al., 2010; Thomas et al., 2010, 2011; Zhang et al., 2002). Long-term sediment transportation degradation causes shoreline erosion and beach retreat (Kundu et al., 2014). Coastal research has been improved by the application of remote sensing and GIS technology (Pramanick et al., 2022).

The current chapters focus on the long-term shoreline modification in covering most part of the Indian Sundarbans using the DSAS tool through remote sensing and GIS applications. The study analyzes the coastline changes or modifications between 2000 and 2020 time period in a part of the Indian Sundarbans using the DSAS tool through the GIS platform. The long-term coastline modification over these 20 years has been estimated by the Linear Regression Rate (LRR) model using the DSAS tool applied by several researchers in the near past (Kaliraj et al., 2015, 2017; Mullick et al., 2020; Bheeroo et al., 2016; Velsamy et al., 2020; Jana et al., 2014; Aladwani, 2022). The innovative and adequate feature of the expected model is the comprehensive analysis of the specific spatial changes over the past 20 years in a large part of the Indian Sundarbans coastal region, which is highly dynamic in nature and also highly impacted by several natural hazards and storms. So, this chapter has been detailing the modification of the coastline on a long-term scale using a statistical model by LRR for four littoral zones of segments. The past research studies have only focused on selective islands of Sundarbans, but this study aims to identify the changes and calculate the erosion-accretion rate covering the maximum areas of the Indian Sundarbans.

4.2 Integrated Method and Data Source

For assessing the coastline modification, satellite images are the most reliable and cost-effective primary data sources. In this current analysis of long-term shoreline modification, Landsat Multi-temporal imagery has been applied, especially Thematic Mapper ™ and Landsat 8 images, to encompass the period of 20 years (2000–2020) in the Indian Sundarbans. The required images were collected from the USGS site on the EarthExplorer website. As the data are cost-free to obtain from the website, they are very adequate for research purposes while maintaining high accuracy across regions. Detailed information about the applied data has been provided in Table 4.1. The present study has selected three multi-resolution images from 2000, 2010, and 2020 for coastline change analysis. After obtaining the images, geo-referencing has been performed for all of them.

4.2.1 Shoreline Delineation

Each band of satellite images has been converted from DN values (Digital Number) to reflectance and radiance values after radiometric correction of the utilized TM and Landsat 8 images (Suyarso et al., 2023). Automatic coastline delineation is very complicated in cases of places where the land–water boundary has saturated due to the water (Jangir et al., 2016; Maiti & Bhattacharya, 2009; Saravanan et al., 2019). Due to this complication, the tasseled cap modification method is more reliable for delineating the coastline in case of TM and Landsat 8 images (Patel et al., 2021; Cenci et al., 2018). In the case of the TM image, the NIR (Near Infrared) band has been applied for the lowest reflectance of water to draw the outline of the coastline. The average ratio has been investigated between bands 2 and 4 and bands 2 and 5 for eliminating the vegetation from the coastline. The water reflected by the combination of 5 in band 2 and various vegetation areas is designated as water indices. The binary data has been converted into raster to vector data, and the coastline boundary is prominently delineated.

Table 4.1 Data applied for the present study

Year	Sensor	Resolution (m)	Path/row	Date	Projection	Datum	Source
2000	Landsat TM	30	138/45	11/9/2000	UTM 45	WGS84	USGS
2010	Landsat TM	30	138/45	12/23/2010	UTM 45	WGS84	USGS
2020	OLI/TIRS	30	138/45	12/18/2020	UTM 45	WGS84	USGS

4.2.2 Method of Casting Transect and Determination of Shoreline Modification

Primarily, a single baseline has been drawn from the coastline before generating the transect on the statistics through either a specific distance or buffer method from the beach (Crowell et al., 1997a, b; Nandi et al., 2016). This method is applicable due to the same sinuosity condition as the nearest beach and buffer method, which is the most perfect process to demarcate the baseline (Nassar et al., 2019). To draw the baseline, a few databases have been required, i.e., SHAPE Length, OBJECTID, GROUP, ID, SHAPE, and CastDir, Offshore in the DSAS analysis system. In the analysis, a 1000 m buffer has been selected for littoral zone 1 and 3000 m for zone 2, and 2000 and 750 m for zones 3 and 4, respectively. After that, transect were laid over baseline as an orthogonal direction with spacing 50 m. to identify the modification rate of shoreline, DSAS is most imperative extension of Arc GIS platform applying various statistical method, i.e., End Point Rate (EPR), Net Shoreline Movement (NSM), Shoreline Change Envelope (SCE) and Linear Regression Rate (LRR) (Abou Samra and Ali, 2021; Den Boer & Oele, 2018; Hossen & Sultana, 2023; Akhter et al., 2024, Murray et al., 2023; Thieler et al., 2005, 2009). For the current long-term shoreline modification analysis, the LRR statistical method has been applied within four littoral zones. In the short-term change analysis chapters, the littoral zone selection and detailed information, as well as the uncertainty, are also described in detail. The short-term analysis has been performed by the EPR method using two successive shorelines because the advantage of this method is that only two data points are required, and the other is neglected. The disadvantage of this EPR model is that the statistics cannot compute more than two datasets with their attributes (Dolan et al., 1991; Crowell et al., 1991; Dike et al., 2024; Mukhopadhyay et al., 2018). For the long-term analysis of 20 years in the present research, the Linear Regression Rate (LRR) method (Nath et al., 2021a, b, 2024a; Abd-Elhamid et al., 2023; Kılar, 2023; Mujabar & Chandrasekar, 2013). In this method, LRR can be computed using the least square regression method, where all shoreline points have been calculated in a single transect, and no shoreline point is neglected for the analysis. To calculate and estimate the shoreline mediation rate over the study area, an erosion-accretion pattern has been observed through the obtained values of each transect using the LRR model. In the transect model, the regression of shoreline means seaward movement of the transect from baseline indicates the accretion or positive changes of shoreline. On the other hand, the transgression of the shoreline or landward movement of the transect from baseline demarked the erosion or negative changes of the shoreline, defining the erosion. To calculate the change rate, a transect spacing of 50 m was chosen. In the current research work, four littoral zones have been analyzed by the LRR method, where the first zone has been calculated for a length of 598.64 km, covering the maximum part of the Indian Sundarbans. The second zone has been covered only 22.86 km, a small island, the third zone covers 75 km, and four zone indicates the 25.07 km shoreline.

4.3 Result and Discussion

To conduct the coastline modification from 2000 to 2020, an extensive analysis has been performed using the LRR method by DSAS. Using the least square regression model, the change rate was calculated for all coastline positions of each transect, starting with the oldest to the most recent one. The obtained negative values from the transect represent the erosion, while the positive values from the transect mark the accretion of shoreline movement. The values have been determined by the color that represents the classes of erosion-accretion based on the values of the transect. Here, five distinct classes have been chosen based on values, i.e., high erosion zone (>−40 m), moderate erosion zone (−39 to −25 m), low erosion zone (−24 to 0), low accretion zone (0–25 m), and high accretion zone (>25). All the littoral zones were classified with these five specific erosion and accretion classes.

4.3.1 Long-Term Shoreline Modification Condition of Littoral Zone 1

The total 598.64 km length of shoreline has been analyzed for segment or littoral zone (LZ) 1. In LZ 1, a total 3945 number of transects have been generated to calculate the LRR from the baseline. A 1000 m buffer distance has been chosen from the baseline with a spacing of 50 m. In the analysis, 378 transects (9.58%) have depicted the erosional pattern over the zone. On the other hand, 3211 number of transects portray the accretion pattern over the region. The entire region has been observed under low accretion conditions. The eastern part of the region depicts the high accretion pattern depicted by Fig. 4.1. Mostly, the entire region is covered by the two types of patterns, i.e., low erosion and low accretion. The change rate of shoreline varies between 3.47 and −10.62 m in LRR values. The high erosion zone depicted near Lalganj Beach and Lothian Island is an erosion range that varies between −41.15 and −44.5 m. The moderate erosion rate of the region has fluctuated between −27.35 and 36.8 m. A graph has been prepared based on the obtained LRR value to depict the erosion-accretion condition of the region (Fig. 4.2). Kalinagar, Minnon Island, Baliara, Bakkhali, Lalaganj Beach, Tater Bazar Ferry, Netidhopani Ghat, Jharkhali Beach have been marked as a low erosional regime. During these 20 years, the Progradation and regression of the shoreline have been gradually changing, and the low accretion zone has been converted into a low erosional regime over time period. The shoreline under marginal erosion needs to be considered under observation to protect it from further erosion.

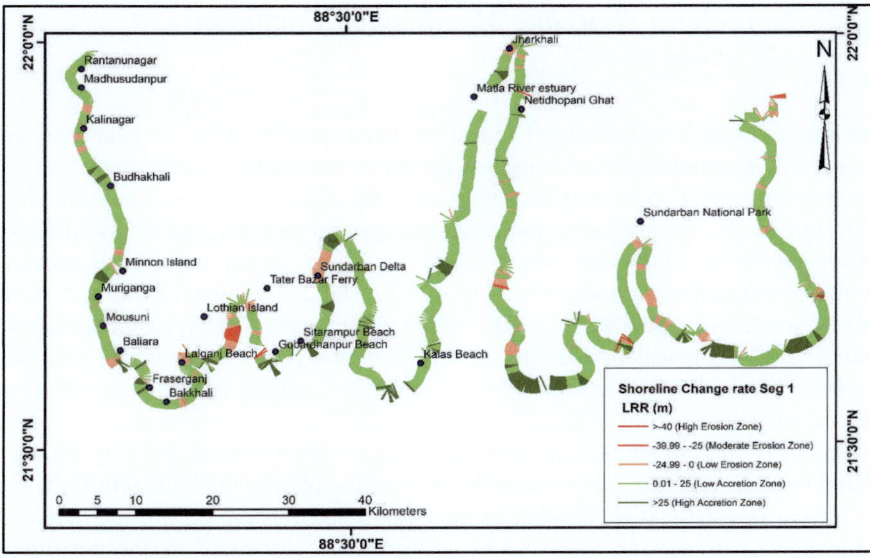

Fig. 4.1 Erosion-accretion zone of segment 1

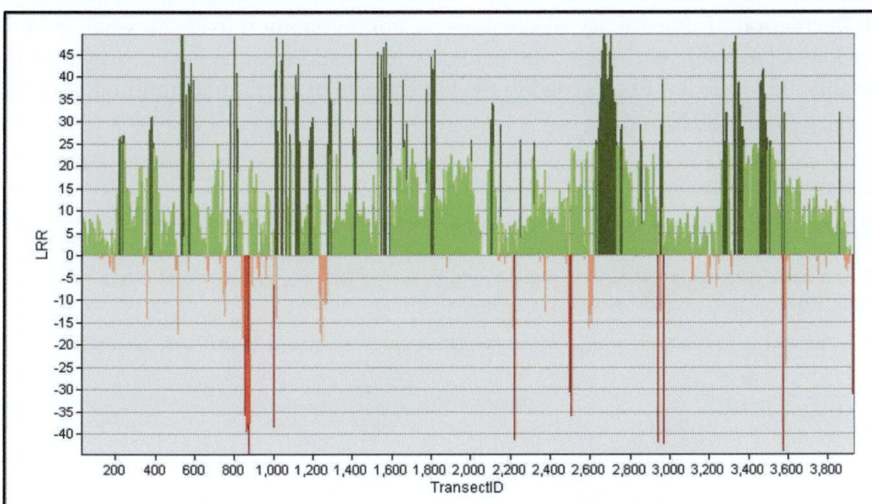

Fig. 4.2 Graph of Erosion-accretion rate by LRR, segment 1

4.3.2 Long-Term Shoreline Modification Condition of Littoral Zone 2

The long-term analysis for littoral zone 2 has been considered for the island named Bangaduni, is a small 22.86 km length of shoreline. The island has been depicting the dynamic movement and versatile changes over the region (Fig. 4.3). The island has been showing the Progradation and regression pattern over the island. Five distinct classes have been prominently observed on this island. Based on the LRR value, a graphical representation is shown in Fig. 4.4. A total 277 number of transects have been generated to calculate the LRR, where 37 transects displayed the erosional pattern, and 240 showed the accretion pattern in the analysis. To generate the transect, a 3000 m buffer distance has been chosen with 50 m spacing of each transect from the baseline.

The lowest part or southern part of the island has been encompassed by the high accretion zone. The maximum accretion rate was found to be 96.51 m, which means 4.8 m/year in LRR statistics. Most of the eastern and western part of the depicted the low accretion pattern, range varies between 7.72 and 22.57 m. the maximum high to moderate erosion zone was occupied over the northern part of the island where the highest erosion rate was reaching out as −63.76 m. the moderate to low erosion rate was varies between −1.66 and 32.33 m. the graph also clearly displayed the fluctuation of dynamicity of erosion-accretion condition over the small island. The low accretion patterns have the tendency to convert into the low erosional zone. Notably, this zone has been showing the highest rate of erosion-accretion during the 20 years (2000–2020).

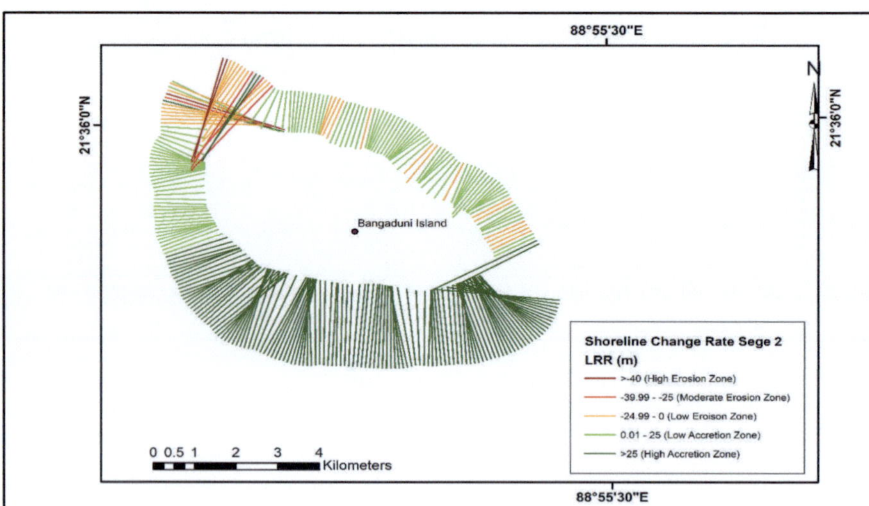

Fig. 4.3 Erosion-accretion zone of segment 2

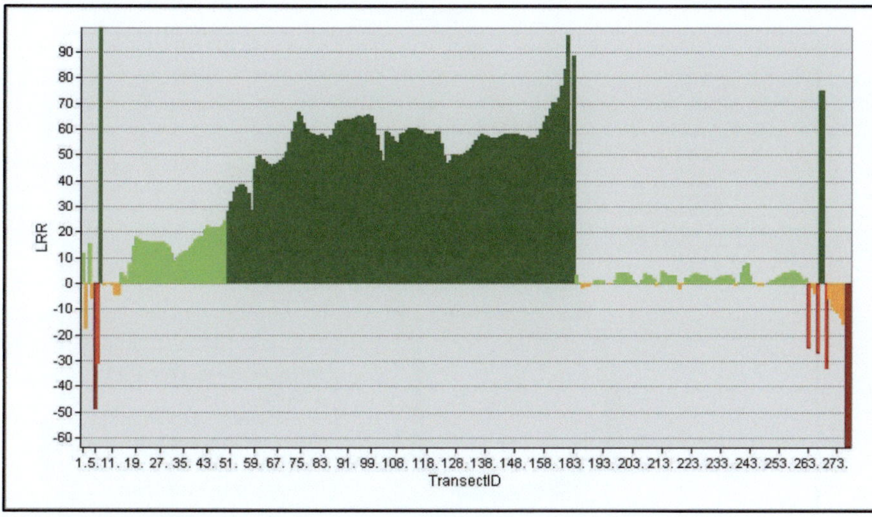

Fig. 4.4 Graph of Erosion-accretion rate by LRR, segment 2

4.3.3 Long-Term Shoreline Modification Condition of Littoral Zone 3

The littoral zone 3 covers the most significant island of the Indian Sundarbans, named Sagar Island. The length of the 75 km shoreline of Sagar Island shows the massive fluctuation of this time period, depicted by Fig. 4.5. To calculate the LRR for Sagar Island shoreline modification rate, 713 transects have been used with a 750 m buffer distance, spacing 50 m. 219 number means 30.72% of the transects have been employed to portray the erosional pattern over segment three, where 69.28% of the transects were involved for showing the accretion pattern. The northern part of the segment has revealed a low to moderate erosion pattern. The highest erosion rate was observed at −60.79 m of LRR values. The average rate of erosion has varied between −1.37 and 37.9 m overall along the shoreline. The highest accretion has been depicted by the southern part of the island, with only small patches of high accretion noticed. The maximum shoreline rate is 31.69 m obtained by the LRR value. A small patch area has been revealed with high erosion in the Sagar Island, which is −60.73 m in the northern part. The overall area has fluctuated between a low erosion and a low accretion zone, but Sagar Island revealed the maximum changes during these 20 years. The island rate of erosion-accretion analysis has been depicting that the shoreline needs to be considered for protecting from future erosion patterns. Figure 4.6 has been prepared for the graphical representation of the LRR value obtained by the analysis in Sagar Island. The prominent erosion-accretion condition has been observed through the graph.

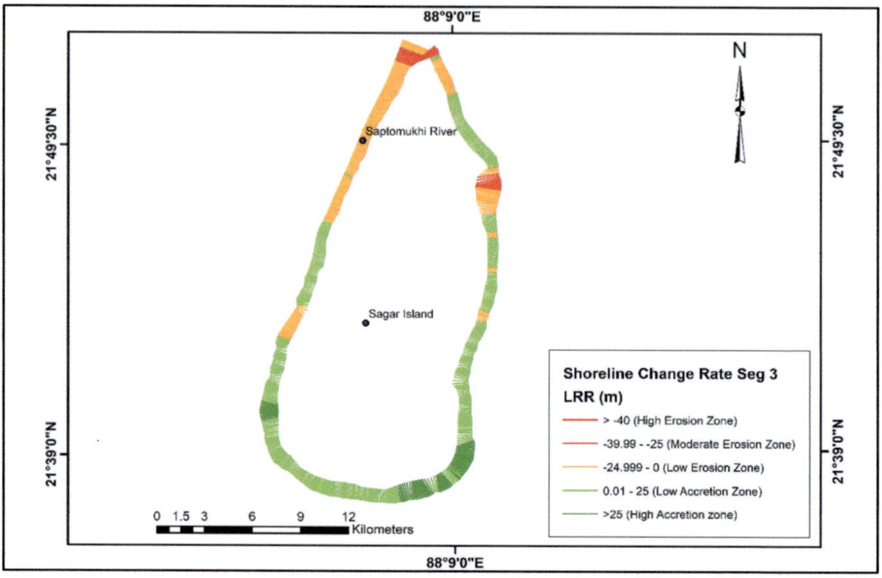

Fig. 4.5 Erosion-accretion zone of segment 3

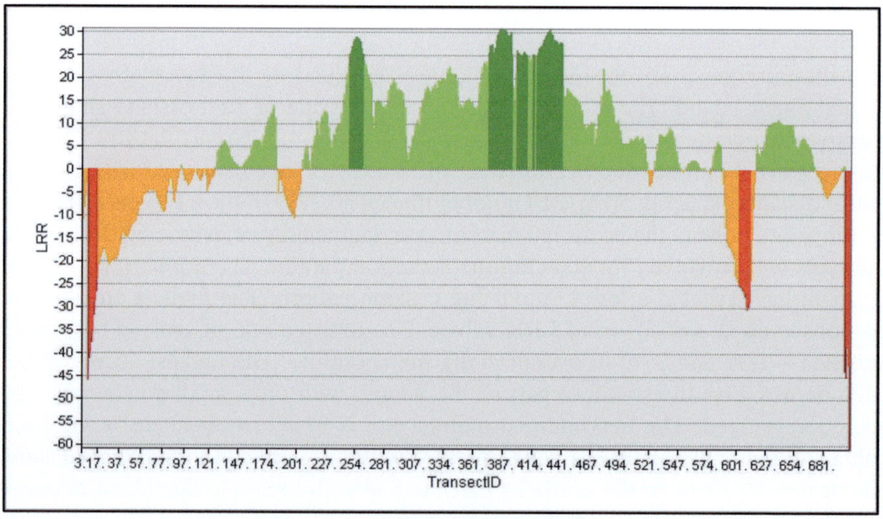

Fig. 4.6 Graph of Erosion-accretion rate by LRR, segment 3

4.3.4 Long-Term Shoreline Modification Condition of Littoral Zone 4

The segment or littoral zone 4 has been showing a small island named Murmu Island in the part of the Indian Sundarban. The total 25.07 km length of the island has been analyzed by the LRR statistical method using 238 transects. Only 20 transects have revealed the erosional pattern over the island. This island has been showing the concentration of accretion or transgression of shoreline movement or seaward movement, indicating the positive changes of shoreline (Fig. 4.7). After the LRR analysis, the obtained data has been depicted through a graphical representation where accretion and erosion prominently measure (Fig. 4.8). The change rate in average varies between 2.14 and 53.75 m. The northern part of the island has been marked only as a low erosion pattern, where the change rate observed is −1.67 to −9.84 m. The highest accretion zone has been covered in the southern part of the island; the high rate of accretion by LRR is 60.34 m. The entire island has been divided into three zonation classes, i.e., low erosion, low accretion, and high accretion zones (Fig. 4.7). This island is marked as under very positive sediment formation during these 20 years (2000–2020). After an entire analysis of part of the Indian Sundarbans, this Murmu Island shows very positive changes over this time period, which is an indication of the significant sediment transportation of the island.

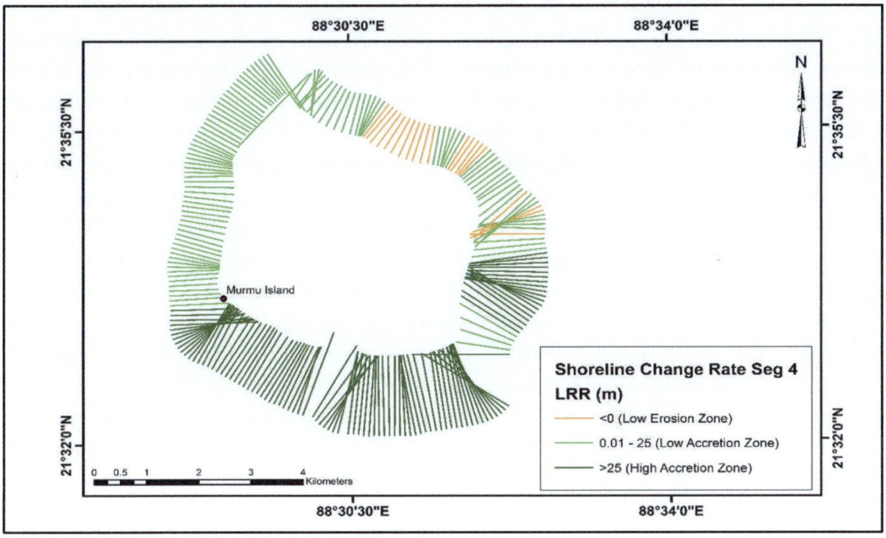

Fig. 4.7 Erosion-accretion zone of segment 4

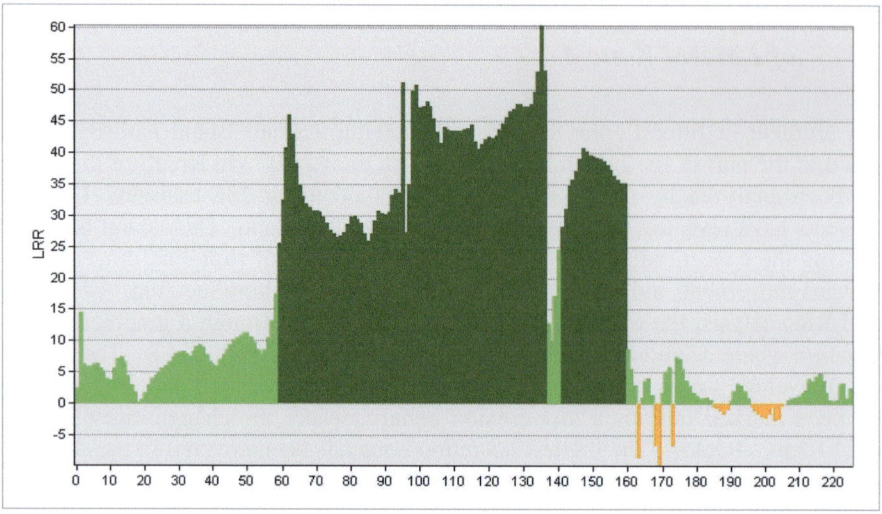

Fig. 4.8 Graph of Erosion-accretion rate by LRR, segment 4

4.4 Discussion

The importance of coastal erosion, accretion, and shoreline modification on the Indian part of the Sundarbans has been emphasized in the present chapter. The analysis of each littoral zone has depicted the scenario of erosion and accretion patterns over a 20-year time period on a long-term scale. The present chapter emphasizes the shoreline change rate by the DSAS analysis system through the GIS platform using the LRR statistical method, where the negative and positive obtained values indicate erosion and accretion. The period from 2000 to 2020 has been selected for the current long-term analysis. Some major storm surges also occurred during this time period, i.e., Aila, Amphan, etc. Most of the coastline has been covered in segment 1, and the other segments are divided into small islands. The past chapter already showed the short-term change of shoreline over the time period, dividing this 20-year period into two time scales. The coastline has been found under substantial changes over all segments, but the largest modification has been taking place in segment 3 or Sagar Island. Segment 2 also shows evidence of an erosional pattern, but a prominent transgression has been revealed over segment 4. Except for a few portions of shoreline, the entire segment 4 is demarked as a major concentration of sedimentation over the region. But the first segment shows a few parts under a moderate to high erosional pattern, which needs to be considered in the near future. Most of the estuary areas have been observed under erosional conditions. The island shape in Sundarbans has been continuously modified by the interface of human activities, natural hazards, and storms. The islands once had an abundant supply of fresh water, but due to geological accessibility, the Hooghly River shifted eastwards, resulting in a significant reduction in flow and several losses due to storage and siltation of the navigable channels. The major erosional changes have been demarked in Sagar Island; the northern part of the island has experienced moderate to high erosion. The LULC change analysis has also revealed the enhancement of population and settlement in Sagar Island, which also emphasizes the erosional tendency over the region. Enhancements of vegetation cover in all parts of the Sundarbans coastline can reduce the erosional tendency, such as mangroves, bamboo, and caesalpinia. Even if there is forested vegetation along the riverbanks or naturally occurring in the riverbed, floodwaters still overflow. Affected riverine areas, islands, and vegetated areas may experience coastal erosion. The outcome of the present research work can be utilized in various planning, resilience, development, or infrastructural implementation. Some other studies have done their research on various parts of the Indian coastline modification and also resulted in the coastal erosion and hazard scenarios in long- and short-term scales (Mondal et al., 2021; Nath et al., 2021a, b, c, 2023, 2024a, b; Jangir et al., 2016; Baral et al., 2018; Kar et al., 2021; Thomas et al., 2023) using LRR and EPR model by DSAS.

4.5 Conclusion

To complete the long-term shoreline modification over part of the Indian Sundarbans, the present work applied the Landsat 8 and TM satellite images collected from the USGS site. The net changes of the chosen area have been computed by the DSAS analysis technique. The shoreline of the present area has been rapidly fluctuating over the past 20 years. A minor erosion has been remarked in only segment 4. Other segments have been encompassed by the low to moderate erosion. But a positive change has been noticed that segment 2 and 4 accumulated the major concentration of high accretion pattern by the LRR model. On the other hand, the fluctuating dynamic movement has also been found in segment 2 and 3. The shape of Bagaduni Island is continuously changed by the erosion of the coastline. In segment 1 most of the tourist-attracted beaches are depicted by the moderate to high erosional pattern, i.e., Bakkhali, Lalganj, Lothian, Kalas, Jharkhali, etc. Additionally, examination into the subject reveals that natural factors such as tide, wave action, and currents cause coastal erosion, and man-made activities also influence the changes in the shoreline adversely.

References

Abd-Elhamid, H. F., Zeleňáková, M., Barańczuk, J., Gergelova, M. B., & Mahdy, M. (2023). Historical trend analysis and forecasting of shoreline change at the Nile Delta using RS data and GIS with the DSAS tool. *Remote Sensing, 15*(7), 1737. https://doi.org/10.3390/rs15071737.

Abou Samra, R. M., & Ali, R. R. (2021). Applying DSAS tool to detect coastal changes along Nile Delta, Egypt. *The Egyptian Journal of Remote Sensing and Space Science, 24*(3), 463–470. https://doi.org/10.1016/j.ejrs.2020.11.002.

Akhter, F., Hoque, M. E., & Xu, N. (2024). Geospatial analysis of shoreline and areal dynamics in the Ganges deltaic island of Bangladesh using the GIS-DSAS technique. *Regional Studies in Marine Science, 73*, Article 103495. https://doi.org/10.1016/j.rsma.2024.103495.

Aladwani, N. S. (2022). Shoreline change rate dynamics analysis and prediction of future positions using satellite imagery for the southern coast of Kuwait: A case study. *Oceanologia, 64*(3), 417–432.

Apostolopoulos, D., & Nikolakopoulos, K. (2021). A review and meta-analysis of remote sensing data, GIS methods, materials and indices used for monitoring the coastline evolution over the last twenty years. *European Journal of Remote Sensing, 54*, 240–265. https://doi.org/10.1080/22797254.2021.1904293.

Apostolopoulos, D. N., Avramidis, P., & Nikolakopoulos, K. G. (2022). Estimating quantitative morphometric parameters and spatiotemporal evolution of the Prokopos Lagoon using remote sensing techniques. *Journal of Marine Science and Engineering, 10*(7), 931. https://doi.org/10.3390/jmse10070931.

Baig, M. R. I., Ahmad, I. A., Shahfahad, Tayyab, M., & Rahman, A. (2020). Analysis of shoreline changes in Vishakhapatnam coastal tract of Andhra Pradesh, India: An application of digital shoreline analysis system (DSAS). *Annals of GIS, 26*(4), 361–376. https://doi.org/10.1080/19475683.2020.1815839.

Baral, R., Pradhan, S., Samal, R. N., & Mishra, S. K. (2018). Shoreline change analysis at Chilika Lagoon Coast, India using digital shoreline analysis system. *Journal of the Indian Society of Remote Sensing, 46*(10), 1637–1644. https://doi.org/10.1007/s12524-018-0818-7.

Bheeroo, R.A., Chandrasekar, N., Kaliraj, S. et al. (2016). Shoreline change rate and erosion risk assessment along the Trou Aux Biches–Mont Choisy beach on the northwest coast of Mauritius using GIS-DSAS technique. *Environmental Earth Sciences, 75*, 444. https://doi.org/10.1007/s12665-016-5311-4.

Boak, E. H., & Turner, I. L. (2005a). Shoreline definition and detection: A review. *Journal of Coastal Research, 21*(4), 688–703. https://doi.org/10.2112/03-0071.1.

Burningham, H., & French, J. (2017). Understanding coastal change using shoreline trend analysis supported by cluster-based segmentation. *Geomorphology, 282*, 131–149. https://doi.org/10.1016/j.geomorph.2016.12.029.

Cenci, L., Disperati, L., Persichillo, M. G., Oliveira, E. R., Alves, F. L., & Phillips, M. (2018). Integrating remote sensing and GIS techniques for monitoring and modeling shoreline evolution to support coastal risk management. *GIScience & Remote Sensing, 55*(3), 355–375. https://doi.org/10.1080/15481603.2017.1376370.

Chowdhury, P., Lakku, N. K. G., Lincoln, S., Seelam, J. K., & Behera, M. R. (2023). Climate change and coastal morphodynamics: Interactions on regional scales. *Science of the Total Environment, 899*, Article 166432. https://doi.org/10.1016/j.ocecoaman.2022.106037.

Ciampalini, A., Consoloni, I., Salvatici, T., Di Traglia, F., Fidolini, F., Sarti, G., & Moretti, S. (2015). Characterization of coastal environment by means of hyper-and multispectral techniques. *Applied Geography, 57*, 120–132. https://doi.org/10.1016/j.apgeog.2014.12.024.

Crowell, M., Douglas, B. C., & Leatherman, S. P. (1997a). On forecasting future US shoreline positions: A test of algorithms. *Journal of Coastal Research, 13*(4), 1245–1255.

Crowell, M., Douglas, B. C., & Leatherman, S. P. (1997). On forecasting future US shoreline positions: A test of algorithms. *Journal of Coastal Research*, 1245–1255.

Dasgupta, S., Laplante, B., Meisner, C., Wheeler, D., & Yan, J. (2009). The impact of sea level rise on developing countries: A comparative analysis. *Climatic Change, 93*(3–4), 379–388. https://doi.org/10.1007/s10584-008-9499-5.

Den Boer, E. L., & Oele, A. C. (2018). Determination of shoreline change along the East-Java coast, using Digital Shoreline Analysis System. In *MATEC Web of Conferences* (Vol. 177, p. 01022). EDP Sciences.

Dev, S. D., Deepchand, V., Anoop, M. S., Krishnaprasad, P. K., Nazeer, M. N., Singh, Y., Arjun, S., & Prasanth, R. S. (2023). Influence of a fishing harbour on coastal geomorphology of the

southwest coast of India and predictions of its future trends. *Geosystems and Geoenvironment, 2*(2), Article 100179. https://doi.org/10.1016/j.geogeo.2023.100179.
Dike, E. C., Amaechi, C. V., Beddu, S. B., Weje, I. I., Ameme, B. G., Efeovbokhan, O., & Oyetunji, A. K. (2024). Coastal Vulnerability Index sensitivity to shoreline position and coastal elevation parameters in the Niger Delta region, Nigeria. *Science of the Total Environment, 919*, Article 170830.
Dolan, R., Fenster, M. S., & Holme, S. J. (1991). Temporal analysis of shoreline recession and accretion. *Journal of Coastal Research, 7*, 723–744.
Douglas, B. C., & Crowell, M. (2000). Long-term shoreline position predictions and error propagation. *Journal of Coastal Research, 16*(1), 145–152.
Esteves, L. S., Toldo, Jr., E. E., & Dillenburg, S. R. (2002). Long- and short-term coastal erosion in southern Brazil. *Journal of Coastal Research, SI35*, 273–282.
Ghaderi, D., & Rahbani, M. (2020). Detecting shoreline change employing remote sensing images (Case study: Beris Port-East of Chabahar, Iran). *International Journal Of Coastal, Offshore And Environmental Engineering (IJCOE), 5*, 1–8.
Granja, H. M., & Carvalho, G. S. (2000). Inland beach migration (beacherosion) and the coastal zone management (Experienceof the northwest coastal zone of Portugal). *Period Biology, 102*, 413–424.
Hanson, H., Gravens, M. B., & Kraus, N. C. (1988). Prototype applications of a generalized shoreline change numerical model. In *Proceedings of the 21st International Conference on Coastal Engineering, Costa del Sol-Malaga, Spain* (pp. 1265–1279).
Hossain, M. S., Yasir, M., Wang, P., Ullah, S., Jahan, M., Hui, S., et al. (2021). Automatic shoreline extraction and change detection: A study on the southeast coast of Bangladesh. *Marine Geology, 441*, Article 106628. https://doi.org/10.1016/j.margeo.2021.106628.
Hossen, M. F., & Sultana, N. (2023). Shoreline change detection using DSAS technique: Case of Saint Martin Island, Bangladesh. *Remote Sensing Applications: Society and Environment, 30*, Article 100943. https://doi.org/10.1016/j.rsase.2023.100943.
Hui, S., Mengliang, G., Yuliang, G., Mingming, X., Shanwei, L., Yasir, M., et al. (2022). Coastline extraction based on multi-scale segmentation and multi-level inheritance classification. *Frontiers in Marine Science, 9*, 1031417. https://doi.org/10.3389/fmars.2022.1031417.
Jana, A., Biswas, A., Maiti, S., et al. (2014). Shoreline changes in response to sea level rise along Digha Coast, Eastern India: An analytical approach of remote sensing, GIS and statistical techniques. *Journal of Coastal Conservation, 18*, 145–155. https://doi.org/10.1007/s11852-013-0297-5.
Jangir, B., Satyanarayana, A. N. V., Swati, S., Jayaram, C., Chowdary, V. M., & Dadhwal, V. K. (2016). Delineation of spatio-temporal changes of shoreline and geomorphological features of Odisha coast of India using remote sensing and GIS techniques. *Natural Hazards, 82*, 1437–1455. https://doi.org/10.1007/s11069-016-2252-x.
Kaliraj, S., Chandrasekar, N., & Magesh, N. S. (2015). Evaluation of coastal erosion and accretion processes along the southwest coast of Kanyakumari, Tamil Nadu using geospatial techniques. *Arabian Journal of Geosciences, 8*, 239–253. https://doi.org/10.1007/s12517-013-1216-7.
Kaliraj, S., Chandrasekar, N., & Ramachandran, K. K. (2017). Mapping of coastal landforms and volumetric change analysis in the south west coast of Kanyakumari, South India using remote sensing and GIS techniques. *The Egyptian Journal of Remote Sensing and Space Science, 20*(2), 265–282. https://doi.org/10.1016/j.ejrs.2016.12.006.
Kar, P. K., Mohanty, P. K., Pradhan, S., Behera, B., Padhi, S. K., & Mishra, P. (2021). Shoreline change along Odisha coast using statistical and geo-spatial techniques. *Journal of Earth System Science, 130*, 1–20. https://doi.org/10.1007/s12040-021-01703-1.
Kılar, H. (2023). Shoreline change assessment using DSAS technique: A case study on the coast of Meriç Delta (NW Türkiye). *Regional Studies in Marine Science, 57*, Article 102737. https://doi.org/10.1016/j.rsma.2022.102737.

Kundu, S., Mondal, A., Khare, D., Mishra, P. K., & Shukla, R. (2014). Shifting shoreline of Sagar island delta, India. *Journal of Maps, 10*(4), 612–619. https://doi.org/10.1080/17445647.2014.922131.

Kvile, K. Ø., Gundersen, H., Poulsen, R. N., Sample, J. E., Salberg, A. B., Ghareeb, M. E., et al. (2024). Drone and ground-truth data collection, image annotation and machine learning: A protocol for coastal habitat mapping and classification. *MethodsX, 13*, Article 102935. https://doi.org/10.1016/j.mex.2024.102935.

Laignel, B., Vignudelli, S., Almar, R., Becker, M., Bentamy, A., Benveniste, J., & Verpoorter, C. (2023). Observation of the coastal areas, estuaries and deltas from space. *Surveys in Geophysics, 44*(5), 1309–1356. https://doi.org/10.1007/s10712-022-09757-6.

Lakshmi, A., & Rajagopalan, R. (2000). Socio-economic implications of coastal zone degradation and their mitigation: A case study from coastal villages in India. *Ocean & Coastal Management, 43*(8–9), 749–762.

Leatherman, S. P. (2001). Social and economic costs of sea-level rise. In B. C. Douglas, M. S. Kearney, & S. P. Leatherman (Eds.), *Sea-level rise, history and consequences* (pp. 181–223). Academic Press. https://doi.org/10.1016/S0074-6142(01)80011-5.

Mahapatra, M., Ratheesh, R., & Rajawat, A. S. (2014a). Shoreline change analysis along the Coast of South Gujarat, India, using Digital Shoreline Analysis System. *Journal of the Indian Society of Remote Sensing, 42*, 869–876. https://doi.org/10.1007/s12524-013-0334-8.

Mahapatra, M., Ratheesh, R., & Rajawat, A. S. (2014b). Shoreline change analysis along the coast of South Gujarat, India, using Digital Shoreline Analysis System. *Journal of the Indian Society of Remote Sensing, 42*, 869–876.

Maiti, S., & Bhattacharya, A. K. (2009). Shoreline change analysis and its application to prediction: A remote sensing and statistics-based approach. *Marine Geology, 257*(1–4), 11–23. https://doi.org/10.1016/j.margeo.2008.10.006.

Maurya, K., Mahajan, S., & Chaube, N. (2021). Remote sensing techniques: Mapping and monitoring of mangrove ecosystem—A review. *Complex Intell. Syst., 7*, 2797–2818. https://doi.org/10.1007/s40747-021-00457-z.

Mondal, M., Haldar, S., Biswas, A., Mandal, S., Bhattacharya, S., & Paul, S. (2021). Modeling cyclone-induced multi-hazard risk assessment using analytical hierarchical processing and GIS for coastal West Bengal, India. *Regional Studies in Marine Science, 44*, 101779. https://doi.org/10.1016/j.rsma.2021.101779.

Mujabar, P. S., & Chandrasekar, N. (2013). Coastal erosion hazard and vulnerability assessment for southern coastal Tamil Nadu of India by using remote sensing and GIS. *Natural Hazards, 69*, 1295–1314. https://doi.org/10.1007/s11069-011-9962-x.

Mukhopadhyay, A., Ghosh, P., Chanda, A., Ghosh, A., Ghosh, S., Das, S., & Hazra, S. (2018). Threats to coastal communities of Mahanadi delta due to imminent consequences of Erosion–Present and near future. *Science of the Total Environment, 637*, 717–729.

Mullick, M. R. A., Islam, K. A., & Tanim, A. H. (2020). Shoreline change assessment using geospatial tools: A study on the Ganges deltaic coast of Bangladesh. *Earth Science Informatics, 13*, 299–316. https://doi.org/10.1007/s12145-019-00423-x.

Murray, J., Adam, E., Woodborne, S., Miller, D., Xulu, S., & Evans, M. (2023). Monitoring shoreline changes along the southwestern coast of South Africa from 1937 to 2020 using varied remote sensing data and approaches. *Remote Sensing, 15*(2), 317. https://doi.org/10.3390/rs15020317.

Nandi, S., Ghosh, M., Kundu, A., Dutta, D., & Baksi, M. (2016). Shoreline shifting and its prediction using remote sensing and GIS techniques: A case study of Sagar Island, West Bengal (India). *Journal of Coastal Conservation, 20*, 61–80. https://doi.org/10.1007/s11852-015-0418-4.

Nassar, K., Mahmod, W. E., Fath, H., Masria, A., Nadaoka, K., & Negm, A. (2019). Shoreline change detection using DSAS technique: Case of North Sinai coast, Egypt. *Marine Georesources & Geotechnology, 37*(1), 81–95. https://doi.org/10.1080/1064119X.2018.1448912.

Nath, A., Koley, B., Saraswati, S., & Ray, B. C. (2021a). Identification of the coastal hazard zone between the areas of Rasulpur and Subarnarekha estuary, east coast of India using multi-criteria evaluation method. *Modeling Earth Systems and Environment, 7*(4), 2251–2265. https://doi.org/10.1007/s40808-020-00986-5.

Nath, A., Koley, B., Saraswati, S., Choudhury, T., Um, J.-S., & Ray, B. C. (2023). Geospatial analysis of short-term shoreline change behavior between Subarnarekha and Rasulpur estuary, east coast of India using intelligent techniques (DSAS). *GeoJournal, 88*(Suppl 1), 255–275. https://doi.org/10.1007/s10708-022-10683-8.

Nath, A., Koley, B., Saraswati, S., Choudhury, T., Um, J.-S., & Ray, B. C. (2024a). Geo-spatial analysis of erosion accretion dynamics and areal changes along Digha Coastline in West Bengal, India. *Remote Sensing in Earth Systems Sciences, 7*, 183–205. https://doi.org/10.1007/s41976-024-00114-2.

Nath, A., Koley, B., Choudhury, T., Saraswati, S., & Um, J.-S. (2024b). Coastal vulnerability differentiated according to geospatial quartiles method: Rasulpur to Subarnarekha estuary, east coast of India. *Spatial Information Research, 32*, 383–398. https://doi.org/10.1007/s41324-023-00554-w.

Nath, A., Koley, B., Saraswati, S., Bandyopadhyay, K., Bhattacharya, S., & Ray, B. C. (2021b). Evaluation of long-term shoreline changes between Rasulpur and Subarnarekha estuary, east coast of India using remote sensing and GIS techniques. In *Modern Cartography Series* (Vol. 10, pp. 337–351). Academic Press.

Nath, A., Koley, B., Saraswati, S., Bhatta, B., & Ray, B. C. (2021c). Shoreline change and its impact on land use pattern and vice versa–A critical analysis in and around Digha area between 2000 and 2018 using geospatial techniques. *Pertanika Journal of Science & Technology, 29*(1).

Nidhinarangkoon, P., Ritphring, S., Kino, K., & Oki, T. (2023). Shoreline changes from erosion and sea level rise with coastal management in Phuket, Thailand. *Journal of Marine Science and Engineering, 11*, 969. https://doi.org/10.3390/jmse11050969.

Parker, B. B. (2003). The difficulties in measuring a consistently defined shoreline–The problem of vertical referencing. *Journal of Coastal Research, SI38*, 44–56.

Parthasarathy, K. S. S., & Deka, P. C. (2019). Remote sensing and GIS application in assessment of coastal vulnerability and shoreline changes. Changes for the Selangor Coast, Malaysia, using the Digital Shoreline Analysis System technique. *Urban Science, 6*, 71. https://doi.org/10.3390/urbansci6040071.

Patel, K., Jain, R., Patel, A. N., & Kalubarme, M. H. (2021). Shoreline change monitoring for coastal zone management using multi-temporal Landsat data in Mahi River estuary, Gujarat State. *Applied Geomatics, 13*, 333–347. https://doi.org/10.1007/s12518-021-00353-8.

Pramanick, N., Islam, E., Banerjee, S., Mukherjee, R., Maity, A., Acharyya, R., et al. (2022). Threats from sea level rise and Erosion: A case study of an estuarine inhabited Island Ghoramara, Hooghly Estuary. *Urban Ecology and Global Climate Change*, 321–345. https://doi.org/10.1002/9781119807216.ch16.

Pranzini, E., & Williams, A. T. (2013). *Coastal erosion and protection in Europe* (p. 454). London & NewYork: Earthscan/Routledge.

Rahman, M. K., Crawford, T. W., & Islam, M. S. (2022). Shoreline change analysis along rivers and deltas: A systematic review and bibliometric analysis of the shoreline study literature from 2000 to 2021. *Geosciences, 12*, 410. https://doi.org/10.3390/geosciences12110410.

Rogers, J., Hamer, B., Brampton, A., Challinor, S., Glennester, M., Brenton, P., & Bradbury. (2010). *A Beach management manual* (2nd ed., p. 915). London: CIRIA.

Saddam, M. M. H., Hossain, K. T., Mollick, P., Salauddin, M., Roy, S., Ahmad, A. U., et al. (2025). Shoreline dynamics in response to sea-level changes on the eastern coast of Bangladesh. *Environmental Monitoring and Assessment, 197*, 942. https://doi.org/10.1007/s10661-025-14375-x.

Saravanan, S., Parthasarathy, K. S. S., & Vishnuprasath, S. R. (2019). Monitoring spatial and temporal scales of shoreline changes in the Cuddalore Region, India. In *Coastal zone management* (pp. 99–112). Elsevier. https://doi.org/10.1016/B978-0-12-814350-6.00004-5.

Sekar, L. G., Androws, X., Annaidasan, K., Kumar, A., Kannan, R., Muthusankar, G., & Balasubramani, K. (2024). Assessment of shoreline changes and associated erosion and accretion pattern in coastal watersheds of Tamil Nadu, India. *Natural Hazards Research, 4*(2), 231–238. https://doi.org/10.1016/j.nhres.2023.09.008.

Silva, J. F., & Duck, R. W. (2006). Changing use and hydromorphological adjustment in a coastal lagoon-estuarine system, the Ria de Aveiro, Portugal. *Sediment Dynamics and the Hydromorphology of Fluvial Systems, 306*, 197–204.

Strayer, D. L., & Findlay, S. E. G. (2010). Ecology of freshwater shore zones. *Aquatic Sciences, 72*, 127–163. https://doi.org/10.1007/s00027-010-0128-9.

Suyarso, S., Setiawati, M. D., Supriyadi, I. H., & Prayudha, B. (2023). Climate change indicator, impact, adaptation, and innovation at the local level: Learn from the peoples' experience of the coastal plain of Probolinggo, East Java, Indonesia. In *Climate change, community response and resilience* (pp. 93–118). Elsevier. https://doi.org/10.1016/B978-0-443-18707-0.00005-9.

Tanwari, K., Terefenko, P., Shi, X., Śledziowski, J., & Giza, A. (2025). Coastal zones vulnerability evaluation in the southern Baltic Sea: Shoreline dynamics and land use/land cover changes over five decades. *Science of the Total Environment, 976*, Article 179345. https://doi.org/10.1016/j.scitotenv.2025.179345.

Thieler, E. R., Himmelstoss, E. A, Zichichi, J. L., & Miller, T. L. (2005). The Digital Shoreline Analysis System (DSAS), an ArcGIS extension for calculating historic Shoreline change. https://doi.org/10.3133/ofr20051304.

Thieler, E. R., Himmelstoss, E. A., Zichichi, J. L., & Ergul, A. (2009). The Digital Shoreline Analysis System (DSAS) version 4.0-an ArcGIS extension for calculating shoreline change (No. 2008–1278). US Geological Survey.

Thomas, T., Phillips, M. R., & Williams, A. T. (2010). Mesoscale evolution of a headland bay: Beach rotation process. *Geomorphology, 123*, 129–141.

Thomas, T., Phillips, M. R., Williams, A. T., & Jenkins, R. E. (2011). A multi-century record of linked nearshore and coastal change. *Earth Surface Processes and Landforms, 36*(8), 995–1006. https://doi.org/10.1002/esp.2127.

Thomas, G. A. J., Santha Ravindranath, R. R., Jeyagopal, S., & Thodhal Yoganandham, S. (2023). Statistical analysis of shoreline change reveals erosion and baseline are increasing off the northern Tamil Nadu Coasts of India. *Environmental Monitoring and Assessment, 195*(3), 409. https://doi.org/10.1007/s10661-023-11015-0.

Uçkaç, Ş. (1998). Kıyı Alanlarında Coğrafi Bilgi Sistemlerinin Kullanımı. In Türkiye' nin Kıyı ve Deniz Alanları 2. Ulusal Konferansı, 22–25 Eylül 1998, Ankara, Türkiye Kıyıları 98 Konferansı Bildiriler Kitabı (pp. 557–564).

Velsamy, S., Balasubramaniyan, G., Swaminathan, B., & Kesavan, D. (2020). Multi-decadal shoreline change analysis in coast of Thiruchendur Taluk, Thoothukudi district, Tamil Nadu, India, using remote sensing and DSAS techniques. *Arabian Journal of Geosciences, 13*(17), 838.

Wang, X., He, X., Shi, J., Chen, S., & Niu, Z. (2022). Estimating sea level, wind direction, significant wave height, and wave peak period using a geodetic GNSS receiver. *Remote Sensing of Environment, 279*, Article 113135. https://doi.org/10.1016/j.rse.2022.113135.

Whitfield, A., & Elliott, M. (2011). Ecosystem and biotic classifications of estuaries and coasts. *Treatise on Estuarine and Coastal Science*, 99–124. https://doi.org/10.1016/B978-0-323-90798-9.00108-6.

Woodroffe, C. D. (2002). *Coasts, form, processes and evolution* (p. 623). Cambridge University Press.

Wright, L. D., & Thom, B. G. (2023). Coastal morphodynamics and climate change: A review of recent advances. *Journal of Marine Science and Engineering, 11*(10), 1997. https://doi.org/10.3390/jmse11101997.

Yasir, M., Hui, S., Binghu, H., & Rahman, S. U. (2020). Coastline extraction and land use change analysis using remote sensing (RS) and geographic information system (GIS) technology–A review of the literature. *Reviews on Environmental Health, 35*, 453–460. https://doi.org/10.1515/reveh-2019-0103.

Yasir, M., Hui, S., Hongxia, Z., Hossain, M. S., Fan, H., Zhang, L., et al. (2021). A spatiotemporal change detection analysis of coastline data in Qingdao, East China. *Scientific Programming, 2021*, 1–10. https://doi.org/10.1155/2021/6632450.

Yasir, M., Liu, S., Mingming, X., Wan, J., Pirasteh, S., & Dang, K. B. (2024). ShipGeoNet: SAR image-based geometric feature extraction of ships using convolutional neural networks. *IEEE Transactions on Geoscience and Remote Sensing*. https://doi.org/10.1109/TGRS.2024.3352150.

Zhang, K., Huang, W., & Douglas, B. C. (2002). Shoreline position variability and long-term trend analysis. *Shore Beach, 70*, 31–35.

Zheng, H., Li, X., Wan, J., Xu, M., Liu, S., & Yasir, M. (2023). Automatic coastline extraction based on the improved instantaneous waterline extraction method and correction criteria using SAR imagery. *Sustainability, 15*, 7199. https://doi.org/10.3390/su15097199.

Zorlu, O., & Kusak, L. (2025). An assessment of the long-term change of the Mersin west coastline using digital shoreline analysis system and detection of pattern similarity using fuzzy C-means clustering. *Frontiers in Marine Science, 12*, 1457016. https://doi.org/10.3389/fmars.2025.1457016.

Chapter 5
Shifting Shores and Risk Zones: DSAS-Based Erosion–Accretion Mapping in the Indian Sundarbans

Abstract The coastal region is very dynamic in its environmental nature, which undergoes continuous geomorphological changes by the natural and man-made processes, i.e., waves, tides, storm surges, and climate change impacts. Erosion-accretion vulnerable zonation mapping is a very crucial spatial analysis program where the land loss and gain areas are easily quantified due to the shoreline modification over a certain time period. Multi-temporal satellite data, historical data, and field data are applied to quantify the change rate. Various statistical methods are used to estimate the rate of retreat in shoreline, i.e., Net Shoreline Movement (NSM), End Point Rate (EPR), and Linear Regression Rate (LRR). The present chapter follows the NSM method to find out the erosion-accretion zones in the Indian Sundarbans region with a long-term change rate (2000–2020) time period. The result of the analysis shows that five distinct classes have been classified to show the erosion-accretion hazard zone, i.e., high erosion, moderate erosion, low erosion, low accretion, and high accretion zones, by providing the erosion-accretion hazard zonation mapping as an essential tool for coastal zone management, sustainable development, risk reduction, and mitigation resilience of coastal habitats.

Keywords Shifting Shores · Risk Zones · Digital Shoreline Analysis System (DSAS) · Erosion–Accretion · Net Shoreline Movement (NSM) · Long-term Zonation Mapping · Littoral Zones · Geospatial Techniques · Indian Sundarbans

5.1 Introduction

Shoreline modification is a natural long-term process, and the shape of the shoreline is continuously changing due to erosion-accretion (Kannan et al., 2016; Ranasinghe, 2016). The man-made activities near the coastline have been gradually increasing, which also affects the global coastline. In the current past, the global coastline has witnessed rapid urbanization as well as population growth (Kilibarda et al., 2014). However, the monitoring and management in the fragile coastal zone is essential at the micro level assessment to identify the nature of changes or modification of the

© The Author(s), under exclusive license to Springer Nature Switzerland AG 2026
A. Nath et al., *Shoreline Modification Over the Indian Sundarban*,
SpringerBriefs in GIS, https://doi.org/10.1007/978-3-032-13981-8_5

shoreline, and it also helps to prevent or mitigate the coastal erosion to some extent (Micallef et al., 2018). Coastal zones are modified by several physical and man-made triggering factors, i.e., wave actions, tidal influence, sediment transportation, estuary reshaping, and unplanned urbanization (Basheer Ahammed & Pandey, 2019; Bird, 2011; Carter, 1988; Kaliraj et al., 2017).

Modification in the shoreline is a very natural and dynamic process, so it is required to monitor the coastline regularly (Nayak, 2002). However, to build a proper coastal zone management plan and coastal regulation act it is required to identify the change nature of any coastline with the spatial information as well the land use land cover pattern of particular coastal area with the ecological condition because the sensitivity of coastal region (Camilleri et al., 2017; Khamis et al., 2017; Ramesh et al., 2017).

In the current situation, coastal researchers have been faced with several challenges, including climate change issues and man-made factors, i.e., sea level rise, wave height acceleration, increasing storm frequency, unplanned swells, sediment transportation, lack of surface runoff, saltwater intrusion, groundwater depletion, etc. (Jha et al., 2012; Thampanya et al., 2006). These challenges could be overcome by implementing mitigation strategies to reduce coastal erosion and increase sedimentation near the shoreline through proper assessment of shoreline modification on a long-term scale (Bodin et al., 2017; National Research Council, 2000; Cooper & Zmud, 1990). Now a day, geospatial techniques has reaching out a revolutionary application for assessing the shoreline modification at any coastal region and remote sensing and Geographical Information System is widely used in coastal research due the reliability, precision, coast effective and user-friendly technology with Global Positioning System (GPS) (Blankespoor et al., 2012; Craglia et al., 2012; Shamsi, 2005).

Conventional or traditional methods are still very significant but it takes major time to identify the changes where the GIS techniques are very logistic for the time effective and cost friendly reliability and the precision of technology so the GIS and remote sensing has been widely applied to assess the dynamicity of shoreline that also helpful for coastal zone management and decision-making in coastal zone management (Wright, 2009; Yang, 2009). Thus, the current evolution of data and technologies, as well as the theoretical emphasis of remote sensing and GIS technology, has provided a large-scale benefit to the coastal scientist to identify the erosion-accretion condition through zonation mapping (Burgan & Aksoy, 2018; Chang & Lai, 2014; Ehteram et al., 2018; Nabipour et al., 2020).

Long-term and short-term changes in sea level directly affect the coastline and determine the shape of the coastline position (Gibeaut et al., 2001). Coastal landforms are also modified by the sea level changes, which triggered several coastal hazards. Coastline changes are very sensitive and a major issue for coastal habitats, i.e., mangroves and tidal flats, in the global context. Sundarbans also faces a crisis in India as well as Bangladesh. Current climate models are prepared to assume the condition of sea level rise and erosion-accretion condition in future decades, with the impacts on coastal habitats at lower elevated areas (Canal-Vergés et al., 2025; Oppenheimer et al., 2019; IPCC, 2021; Parthasarathy & Deka, 2019).

In a recent context, coastal erosion is a major concern in countries like India, where extensive population growth has been identified (Mageswaran et al., 2015). The Indian Sundarbans is one of the special regions due to the natural versatility of mangrove and tidal flats that also protect from several storm surges and natural hazards. To delineate the coastline and estimate the modification rate, remote sensing and GIS are the most effective and reliable platforms (Hegde & Akshaya, 2015). In the context of India, several researchers have already applied the erosion-accretion studies through remote sensing and GIS. A study was attempted by Chenthamilselvan et al. (2014), to show the historical shoreline changes information applying the multi-resolution satellite data at the Karnataka coast through the GIS technique. A shoreline proxy model has been built by Kankara et al. (2018) to monitor the coastline changes for the last 2 decades using GIS and satellite images along Andhra Pradesh. An erosion-accretion study was also developed by Natesan et al. (2015) using multi-temporal satellite data through remote sensing and GIS, applying geospatial methods. A Canny edge detection model and image filter process was used by Balasaraswathi and Srinivasalu (2018) to interpret the coastline modification at three specific beaches (Kodiakkarai Harbor, Nagapattinam Beach, and Velankanni Beach).

The most current research work has employed the Digital Shoreline Analysis System (DSAS) to estimate the shoreline position changes worldwide. Based on the change rate, the erosion-accretion zonation map has been prepared along the part of the Indian Sundarbans with various islands. A short-term zonation map was already prepared in the previous chapter by the EPR model through DSAS. But the previous chapter has also depicted the long-term erosion-accretion zonation map by using the Linear Regression model. But the current chapter illustrates the long-term erosion-accretion zonation condition by the most applied and suitable model, named Net Shoreline Movement (NSM). This chapter only focuses on the condition of erosion-accretion over the study region through a zonation mapping using the value obtained by NSM through the GIS platform.

5.2 Methodology for Long-Term Zonation Mapping

The current chapter has been determining to assess the erosion-accretion zonation on a long-term scale during 2000–2020. To complete the analysis of shoreline modification over the region, the research work has been applying the satellite image of Landsat 8 and Thematic Mapper™ of three specific years (2000, 2010, and 2020). The study tries to choose the cloud-free images for analysis.

The 10.8 ArcGIS software has been used to prepare the shoreline dataset. To extract the shoreline, the high tideline has been considered as a proxy of the coastline. The Histogram threshold method was applied to extract the shoreline; the past two chapters have detailed the method. After applying the histogram threshold, the land and water pixels have been separated with the binary number 0 and 1, where 0 is considered as a water pixel and 1 remarked as a land pixel. After the proper classification, the shoreline has been extracted manually. To identify the actual displacement rate of the shoreline, the DSAS tool obtained by the USGS (United States Geological Survey) has been used for transect generation (Oyedotun, 2014). This tool has the ability to analyze the shorelines quickly through the statistical method as an extension of the ArcGIS software to calculate the modification rate (Abou Samra & Ali, 2020). Primarily, a single baseline has been drawn from the coastline before generating the transect on the statistics through either a specific distance or a buffer method from the beach. To draw the baseline, a few databases have been required, i.e., SHAPE Length, OBJECTID, GROUP, ID, SHAPE, and CastDir, Offshore in the DSAS analysis system. In the analysis, a 1000 m buffer has been selected for littoral zone 1 and a 3000 m buffer for zone 2, and 2000 m and 750 m for zones 3 and 4, respectively. After that, transects were laid over the baseline as an orthogonal direction with a spacing of 50 m. to identify the modification rate of the shoreline (Nor et al., 2020). To compute the statistics, the intersection points of each transect from the baseline have been used to compute the displacement rate (Nassar et al., 2019; Velsamy et al., 2020). To calculate the displacement of the shoreline, various statistical methods are available in the DSAS model, but the current work chose the Net Shoreline Movement (NSM) to calculate the rate of change (De Lima et al., 2021). The obtained value by the NSM method has been applied to prepare the zonation map of erosion-accretion during these 20 years.

The model of NSM has been calculated based on the distance of each transect of oldest and youngest shorelines (Dereli and Tercan, 2020). The erosion-accretion zonation map has been prepared for four specific segment or littoral zones. From the baseline to each transect value was acquired as the net shoreline movement of the particular region. The present work has been calculating the long-term rate by taking two shorelines that are 2000 and 2020 for generate the erosion-accretion map. The entire zonation map for each segment has been divided into five zonation classes where high erosion zone remarked as >-300 m, moderate erosion zone as -300 to -150 m, low erosion zone as -150 m to 0, low accretion demarked as 0 to 350 m and high accretion identified by >350 m by NSM value. Here, all the negative values indicate the erosion that means the landward regression and positive values in analysis represent the accretion that obtained by the seaward or transgression of coastline. To showing the erosion-accretion zonation, the studies mostly follows Linear Regression Rate (LRR) and NSM statistical methods. The past chapter showing the short-term zonation by End Point Rate (EPR) as well LRR showing the long-term also. Another prominent method of NSM has been analyzed in the current chapter. The uncertainty has been already calculated in the previous chapter.

$$NSM = Distance\ between\ the\ youngest\ and\ oldest\ shoreline\ (m) \tag{5.1}$$

5.3 Result and Discussion

5.3.1 Erosion-Accretion Zonation Mapping for Segment 1

The coastline modification analysis provides the pattern of erosion-accretion throughout the study region using the DSAS tool. The length of 598.64 km has been analyzed in the littoral zone or segment 1 to show the pattern of erosion-accretion by the NSM model. A total of 3945 transects were generated as orthogonal lines upon the baseline, and the intersection point of each transect was chosen to calculate the NSM. Five classes were used to identify the erosion-accretion zones. Based on the NSM value, all five classes were observed in this region. Based on the obtained value, the map has been prepared. The entire segment has been depicting the low accretion zones to low erosional zones (Fig. 5.1). A 1000 m buffer distance has been chosen from the baseline with a spacing of 50 m. In the analysis, 378 transects (9.58%) have depicted the erosional pattern over the zone. On the other hand, 3211 number of transects portray the accretion pattern over the region. But the region has also displayed high and moderate erosion. Mostly, the tourist-attracted beaches have been depicted as having high to moderate erosional zones. The high erosion rate has varied between −350 m (17.5 m/year) and −837.38 m (41.87 m/year), and the moderate erosional rate has been observed between −167.96 and −237.34 m. The Baliara, Lalganj Beach, Lothian Beach, Tater Bazar Ferry, Sundarbans Delta, Haliday Island Sanctuary, Minnon Island, and Netidhopani have been depicted as moderate to high erosion zones. But the positive sign is that the entire region shows a low to high accretion zone over 20 years. The highest accretion zone has been depicted along the southern part of the region, mostly with the highest rate of 978.54 m. But another concerning matter is that the low accretion zones are susceptible to shifting toward the low erosion zone. For this reason, the entire region needs to be protected from erosional activities.

5.3.2 Erosion-Accretion Zonation Mapping for Segment 2

In the case of the littoral zone, 22.86 km of shoreline has been analyzed using 277 transects. The island has been depicting the dynamic movement and versatile changes over the region (Fig. 5.2). The island has been showing a progradation and regression pattern over time. Five distinct classes have been prominently observed on this island. A total 277 number of transects have been generated to calculate the NSM, where 37 transects displayed the erosional pattern and 240 transects showed the accretion pattern in the analysis. To generate the transect, a 3000 m buffer distance has been chosen with 50 m spacing of each transect from the baseline.

The shoreline of zone 2 depicts some versatility in nature, where the southern part of the shoreline is dominated by a high accretion zone; on the other hand, the northern part reveals a high to moderate erosional zone. The most noticeable factors

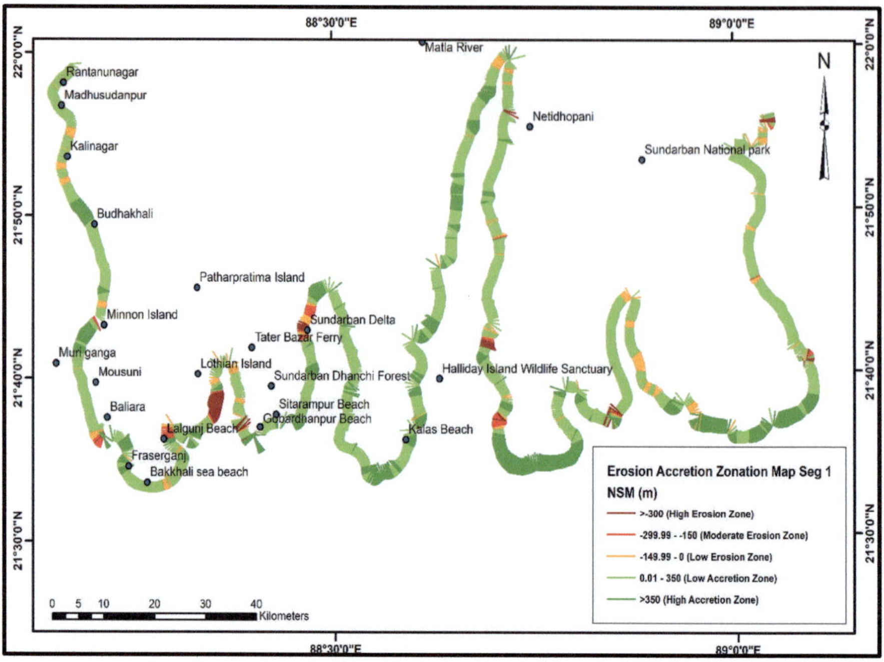

Fig. 5.1 Long-term Erosion-Accretion Zonation map by NSM for segment 1

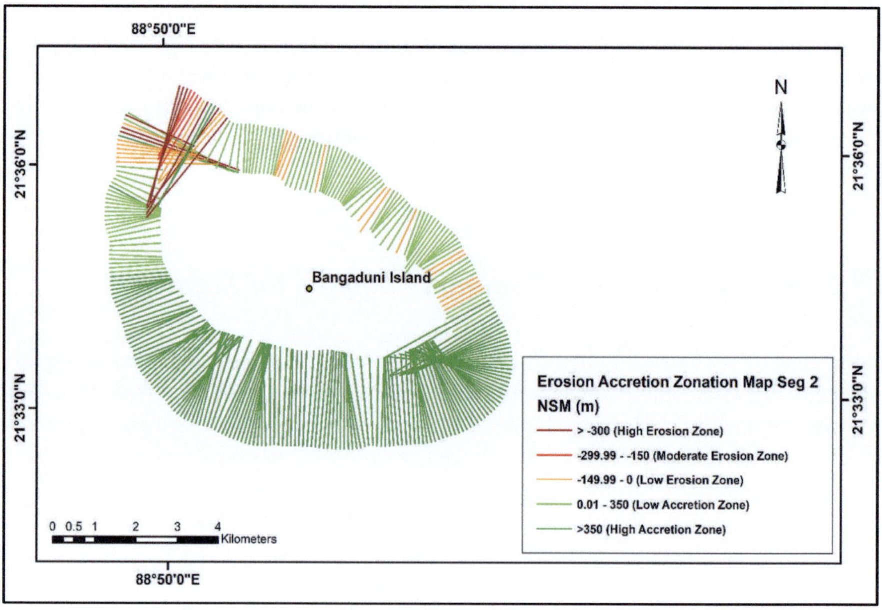

Fig. 5.2 Long-term Erosion-Accretion Zonation map by NSM for segment 2

of this zone are that the estuary part has been mostly dominated by an erosional pattern. The highest rate of accretion was obtained as 1927.77 m (96.38 m/year). The accretion rate varies between 14.63 and 62.77 m/year. The high erosion rate is reaching out at most northern parts with a rate of -64.04 m/year. The low to moderate erosion rate has varied between −2.11 and −33.34 m/year. Notably, this zone has been showing the highest rate of erosion-accretion during the 20 years (2000–2020).

5.3.3 Erosion-Accretion Zonation Mapping for Segment 3

The littoral zone or segment 3 covers a length of 75 km shoreline; the island is named Sagar Island. This island is one of the most significant tourist attractions in the Indian Sundarban. The 20-year-long-term analysis has been depicting the scenario of erosion-accretion zones over the island. To prepare the zonation map by NSM 713 number of transects has been used with a 750 m buffer distance and 50 m spacing. The most important destination, named Gangasagar, has been located in Sagar Island along the Saptamukhi River. The island has been revealed the regression and transgression over the entire shoreline. All the selected erosion-accretion classes have been identified within this segment.

The 219 number means 30.72% of the transects have been employed to portray the erosional pattern over segment three, where 69.28% of the transects were involved for showing the accretion pattern. The northern and western parts of the island have been occupied and dominated by an erosional zone due to the regression of the shoreline. On the other hand, the southern part of the island has been remarkably found under an accretion zone. The river estuary of Saptamukhi has been depicting the high erosional zone for 20 years (Fig. 5.3). The LULC pattern also shows the dynamic changes of Sagar Island in the past chapter. The highest shoreline regression rate has been obtained as −46.23 m/year (−924.58 m), very close to the Saptamukhi River. The average erosion or landward regression rate has varied between −2.68 and −21.27 m/year. The highest transgression rate of the shoreline is 30.95 m/year in the southern part of the island. The Gangasagar shoreline has been dominated by the low accretion zone. The overall area has been fluctuated between low erosion and low accretion zone but Sagar Island revealed the maximum changes during these 20 years. The island rate of erosion-accretion analysis has been depicting that the shoreline needs to be considered for protecting from future erosion patterns.

5.3.4 Erosion-Accretion Zonation Mapping for Segment 3

The segment or littoral zone 4 has been showing a small island named Murmu Island in part of the Indian Sundarban. The total 25.07 km length of the island has been analyzed by the NSM statistical method for the long-term zonation pattern of segment 4 using 238 transects. After the obtained value by NSM, four classes have been

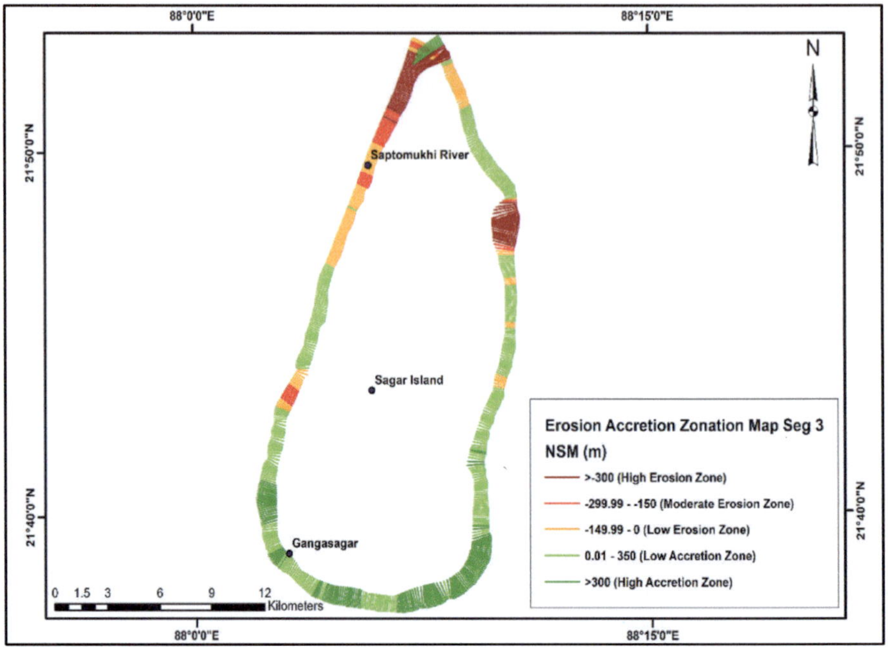

Fig. 5.3 Long-term Erosion-Accretion Zonation map by NSM for segment 3

obtained on the island. Only segment four has not been occupied by the high erosional zone within the entire study region of the Indian Sundarban. Mostly, the Murmu Island shoreline has been dominated by a low to high accretion zone (Fig. 5.4). Only 20 transects have depicted the erosional zone out of 238 transects. This island has been showing the concentration of accretion or transgression of shoreline movement or seaward movement, indicating the positive changes of shoreline (Fig. 5.4). The eastern part of the island has only been focusing on the erosion zone, with the average rate varying between −1.67 and −9.89 m/year. The low to high accretion zone has varied between 2.25 and 60.64 m/year by the shoreline transgression or seaward movement of the shoreline. This island is marked as under very positive sediment formation during these 20 years (2000–2020) of long-term zonation assessment. After an entire analysis of part of the Indian Sundarbans, Murmu Island shows very positive changes over this time period, which indicates the significant sediment transportation of the island.

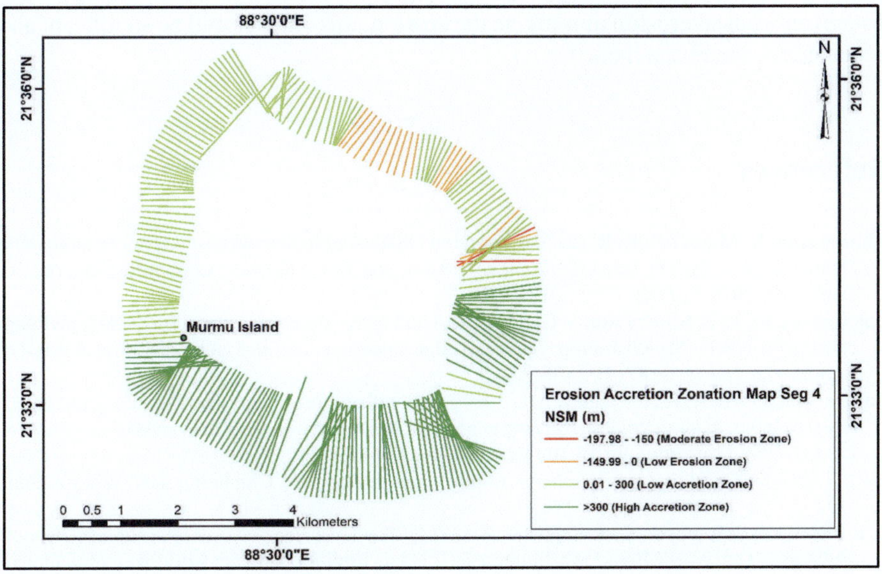

Fig. 5.4 Long-term Erosion-Accretion Zonation map by NSM for segment 4

5.4 Conclusion

To complete the long-term shoreline zonation mapping over part of the Indian Sundarbans, the present work applied the Landsat 8 and TM satellite images collected from the USGS site. The net changes of the chosen area have been computed by the DSAS analysis technique. The chapter has followed the Net Shoreline Movement (NSM) model for reaching out the erosion-accretion zone over a part of the Indian Sundarban. The shoreline of the present area has been rapidly fluctuating over the past 20 years. A minor erosion has been remarked in only segment 4. Other segments have been encompassed by the low to moderate erosion. But a positive change has been noticed that segment 2 and 4 accumulated the major concentration of high accretion pattern by the NSM model. On the other hand, the fluctuating dynamic movement has also been found in segment 2 and 3. The shape of Bagaduni Island is continuously changed by the erosion of the coastline. In segment 1, most of the tourist-attracted beaches are depicted by the moderate to high erosional pattern, i.e., Bakkhali, Lalganj, Lothian, Kalas, Jharkhali, etc. Additionally, examination into the subject reveals that natural factors such as tide, wave action, and currents cause coastal erosion, and man-made activities also influence the changes in the shoreline adversely. The current research work has been followed by two specific statistical methods for assessing the long-term alteration of the coastline in the study area, i.e., NSM and LRR. In this particular chapter, the NSM data has been applied to prepare the zonation map of the study area. By this zonation mapping, the most vulnerable shorelines have been easily and prominently identified, which can turn into hazardous conditions in the near future.

To prevent coastal erosion or mitigate the erosion, this chapter will be significant and adequate for the researchers.

References

Abou Samra, R. M., & Ali, R. R. (2020). Applying DSAS tool to detect coastal changes along Nile Delta, Egypt. *Egyptian Journal of Remote Sensing and Space Science*, xxxx. https://doi.org/10.1016/j.ejrs.2020.11.002.

Balasaraswathi, P., & Srinivasalu, S. (2018). A fast and better shoreline extraction of Nagapattinam coast from TM/ETMþ/OLI images based in edge detection and filter method. *Indian Journal of Geo-Marine Sciences, 47*(9), 1838–1846.

Basheer Ahammed, K. K., & Pandey, A. C. (2019). Shoreline morphology changes along the eastern Coast of India, Andhra Pradesh by using geospatial technology. *Journal of Coastal Conservation, 23*, 331–353. https://doi.org/10.1007/s11852-018-0662-5.

Bird, E. (2011). *Coastal geomorphology: An introduction* (2nd ed.). Chichester, UK: Wiley & Sons Ltd.

Blankespoor, B., Dasgupta, S., & Laplante, B. (2012). *Sea-level rise and coastal wetlands: Impacts and costs*. The World Bank, Washington DC, USA. http://documents.worldbank.org/curated/en/375251468323964212.

Bodin, Ö., Sandström, A., & Crona, B. (2017). Collaborative networks for effective ecosystem-based management: A set of working hypotheses. *Policy Studies Journal, 45*, 289–314. https://doi.org/10.1111/psj.12146.

Burgan, H. I., & Aksoy, H. (2018). Annual flow duration curve model for ungauged basins. *Hydrology Research, 49*, 1684–1695. https://doi.org/10.2166/nh.2018.109.

Camilleri, S., De Giglio, M., Stecchi, F., & Pérez-Hurtado, A. (2017). Land use and land cover change analysis in predominantly man-made coastal wetlands: Towards a methodological framework. *Wetlands Ecology and Management, 25*, 23–43. https://doi.org/10.1007/s11273-016-9500-4.

Canal-Vergés, P., Frederiksen, L., & Egemose, S. (2025). Impacts of sea level rise on Danish Coastal Wetlands–A GIS-based analysis. *Environmental Management, 75*, 1039–1054. https://doi.org/10.1007/s00267-024-02096-9.

Carter, R. W. G. (1988). *Coastal environments: An introduction to the physical*. Academic Press, London, UK.

Chang, F.-J., & Lai, H.-C. (2014). Adaptive neuro-fuzzy inference system for the prediction of monthly shoreline changes in northeastern Taiwan. *Ocean Engineering, 84*, 145–156. https://doi.org/10.1016/j.oceaneng.2014.03.020.

Chenthamilselvan, S., Kankara, R. S., & Rajan, B. (2014). Assessment of shoreline changes along Karnataka coast, India using GIS and remote sensing techniques. *Indian Journal of Geo Marine Sciences, 43*(7), 1286–1291.

Cooper, R. B., & Zmud, R. W. (1990). Information technology implementation research: A technological diffusion approach. *Management Science, 36*, 123–139. https://doi.org/10.1287/mnsc.36.2.123.

Craglia, M., de Bie, K., Jackson, D., Pesarisi, M., Remetey-Fulopp, G., Wang, C., Annoni, A., et al. (2012). Digital earth 2020: Towards the vision for the next decade. *International Journal of Digital Earth, 5*, 4–21. https://doi.org/10.1080/17538947.2011.638500.

Dereli, M.A., Tercan, E. (2020). Assessment of shoreline changes using historical satellite images and geospatial analysis along the Lake Salda in Turkey. *Earth Science Informatics, 13*, 709–718. https://doi.org/10.1007/s12145-020-00460-x.

De Lima, L. T., Fernández-Fernández, S., Espinoza, J. M. D. A., Albuquerque, M. D. G., & Bernardes, C. (2021). End point rate tool for QGIS (EPR4Q): Validation using DSAS and

AMBUR. *ISPRS International Journal of Geo-Information, 10*(3). https://doi.org/10.3390/ijgi10030162.

Ehteram, M., Mousavi, S. F., Karami, H., Farzin, S., Singh, V. P., Chau, K., & El-Shafie, A. (2018). Reservoir operation based on evolutionary algorithms and multi-criteria decision-making under climate change and uncertainty. *Journal of Hydroinformatics, 20*, 332–355. https://doi.org/10.2166/hydro.2018.094.

Gibeaut, J. C., Hepner, T., Waldinger, R., Andrews, J., Gutierrez, R., Tremblay, T. A., & Smyth, R. (2001). Changes in Gulf Shoreline position, Mustang, and North Padre Islands, Texas. A report of the Texas Coastal Coordination Council pursuant to National Oceanic and Atmospheric Administration Award No. NA97OZ0179. GLO Contract Number 00-002R. Prepared by the Bureau of Economic Geology at the University of Texas at Austin. https://oaktrust.library.tamu.edu/server/api/core/bitstreams/1734311f-f2e4-4c4c-b399-ad0d24f69baf/content.

Hegde, A. V., & Akshaya, B. J. (2015). Shoreline transformation study of Karnataka coast: geospatial approach. *Aquatic Procedia, 4*(Icwrcoe), 151–156. https://doi.org/10.1016/j.aqpro.2015.02.021.

IPCC. (2021). Summary for policymakers. In V. Masson-Delmotte, P. Zhai, A. Pirani, S. L. Connors, C. Péan, S. Berger, N. Caud, Y. Chen, L. Goldfarb, M. I. Gomis, M. Huang, K. Leitzell, E. Lonnoy, J. B. R. Matthews, T. K. Maycock, T. Waterfield, O. Yelekçi, R. Yu, & B. Zhou (Eds.), *Climate change 2021: The physical science basis. Contribution of working group I to the sixth assessment report of the intergovernmental panel on climate change* (pp. 3–32). Cambridge, United Kingdom and New York, NY, USA: Cambridge University Press.

Jha, A. K., Bloch, R., & Lamond, J. (2012) *Cities and flooding: A guide to integrated urban flood risk management for the 21st century*. Washington DC, USA: The World Bank.

Kaliraj, S., Chandrasekar, N., & Ramachandran, K. K. (2017). Mapping of coastal landforms and volumetric change analysis in the south west coast of Kanyakumari, South India using remote sensing and GIS techniques. *The Egyptian Journal of Remote Sensing and Space Science, 20*, 265–282. https://doi.org/10.1016/j.ejrs.2016.12.006.

Kankara, R. S., Murthy, R., & Rajeevan, M. (2018). *National assessment of Shoreline Changes along Indian Coast-A status report for 1990–2016*. National Centre for Coastal Research, Chennai.

Kannan, R., Ramanamurthy, M. V., & Kanungo, A. (2016). Shoreline change monitoring in Nellore coast at east coast Andhra Pradesh District using remote sensing and GIS. *Journal of Fisheries and Livestock Production, 4*, 161. https://doi.org/10.4172/2332-2608.1000161.

Khamis, Z. A., Kalliola, R., & Käyhkö, N. (2017). Geographical characterization of the Zanzibar coastal zone and its management perspectives. *Ocean and Coastal Management, 149*, 116–134. https://doi.org/10.1016/j.ocecoaman.2017.10.003.

Kilibarda, Z., Graves, N., Dorton, M., & Dorton, R. (2014). Changes in beach gravel lithology caused by anthropogenic activities along the southern coast of Lake Michigan, USA. *Environmental Earth Sciences, 71*, 1249–1266. https://doi.org/10.1007/s12665-013-2529-2.

Mageswaran, T., Mohan, R. V., Selvan, C. S., Arumugam, T., Usha, T., & Kankara, R. S. (2015). Assessment of shoreline changes along Nagapattinam coast using geospatial techniques. *Int. J. Geomatics Geosci., 5*(4), 555–563.

Micallef, S., Micallef, A., & Galdies, C. (2018). Application of the coastal hazard wheel to assess erosion on the Maltese coast. *Ocean & Coastal Management, 156*, 209–222. https://doi.org/10.1016/j.ocecoaman.2017.06.005.

Nabipour, N., Mosavi, A., Hajnal, E., Nadai, L., Shamshirband, S., & Chau, K.-W. (2020). Modeling climate change impact on wind power resources using adaptive neuro-fuzzy inference system. *Engineering Applications of Computational Fluid Mechanics, 14*, 491–506.

Nassar, K., Mahmod, W. E., Fath, H., Masria, A., Nadaoka, K., & Negm, A. (2019). Shoreline change detection using DSAS technique: Case of North Sinai coast, Egypt. *Marine Georesources & Geotechnology, 37*(1), 81–95. https://doi.org/10.1080/1064119X.2018.1448912.

Natesan, U., Parthasarathy, A., Vishnunath, R., Kumar, G. E. J., & Ferrer, A. (2015). Monitoring long term shoreline changes along Tamil Nadu, India using geospatial techniques. *Aquatic Procedia, 4*(Icwrcoe), 325–332. https://doi.org/10.1016/j.aqpro.2015.02.044.

National Research Council. (2000). Resources Commission on Geosciences Environment, Water Science and Technology Board, Ocean Studies Board & Committee on the Causes and Management of Coastal Eutrophication 2000 Clean Coastal Waters: Understanding and Reducing the Effects of Nutrient Pollution. Washington DC, USA: National Academies Press.

Nayak, S. (2002). Use of satellite data in coastal mapping. *Indian Cartographer, 22*, 147–157.

Nor, N. A., Tahar, K. N., Suprijo, T., & Sulaiman, S. A. (2020). Shoreline changes analysis along the Coast of Kuala Terengganu, Malaysia using DSAS. In: *2020 11th IEEE Control and System Graduate Research Colloquium (ICSGRC)* (pp. 276–281).

Oppenheimer, M., Glavovic, B.C., Hinkel, J., van de Wal, A. K. R., Magnan, A., Abd-Elgawad, R., et al. (2019). Sea level rise and implications for low-lying islands, coasts and communities. In H. -O. Pörtner, D. C. Roberts, V. Masson-Delmotte, P. Zhai, M. Tignor, E. Poloczanska, K. Mintenbeck, A. Alegría, M. Nicolai, A. Okem, J. Petzold, B. Rama, & N. M. Weyer (Eds.), *IPCC special report on the ocean and cryosphere in a changing climate* (pp. 321–445). Cambridge, UK and New York, NY, USA: Cambridge University Press. https://doi.org/10.1017/9781009157964.006.

Oyedotun, T. D. T. (2014). *Shoreline geometry: DSAS as a tool for historical trend analysis, geomorphological techniques* (Vol. 2, pp. 1–12). British Society for Geomorphology.

Parthasarathy, K. S. S., & Deka, P. C. (2019). Remote sensing and GIS application in assessment of coastal vulnerability and shoreline changes: A review. *ISH Journal of Hydraulic Engineering, 113*. https://doi.org/10.1080/09715010.2019.1603086.

Ramesh, R., Purvaja, R., Krishnan, P. & Lakshmi, A. (2017). Conservation of coastal wetlands: An appraisal of the policy and legal framework in South Asian nations. In B. Anjan Kumar Prusty, R. Chandra, & P. A. Azeez (Eds.), *Wetland science* (pp. 515–544). Switzerland: Springer Nature.

Ranasinghe, R. (2016). Assessing climate change impacts on open sandy coasts: A review. *Earth-Science Reviews, 160*, 320–332. https://doi.org/10.1016/j.earscirev.2016.07.011.

Shamsi, U. M. (2005). *GIS applications for water, wastewater, and stormwater systems*. Boca Raton, FL, USA: CRC Press. https://uodiyala.edu.iq/uploads/PDF%20ELIBRARY%20UODIYALA/EL32/GIS%20Applications%20for%20Water,%20Wastewater.pdf.

Thampanya, U., Vermaat, J. E., Sinsakul, S., & Panapitukkul, N. (2006). Coastal erosion and mangrove progradation of Southern Thailand. *Estuarine, Coastal and Shelf Science, 68*, 75–85.

Velsamy, S., Balasubramaniyan, G., Swaminathan, B., & Kesavan, D. (2020). Multi-decadal shoreline change analysis in coast of Thiruchendur Taluk, Thoothukudi district, Tamil Nadu, India, using remote sensing and DSAS techniques. *Arabian Journal of Geosciences, 13*(17), 1–12. https://doi.org/10.1007/s12517-020-05800-1.

Wright, D. J. (2009). Spatial data infrastructures for coastal environments. In X. Yang (Ed.), *Remote sensing and geospatial technologies for coastal ecosystem assessment and management* (pp. 91–112). Heidelberg, Germany: Springer-Verlag.

Yang, X. (2009). *Remote sensing and geospatial technologies for coastal ecosystem assessment and management*. Berlin, Germany: Springer-Verlag.

Chapter 6
Decoding Landscape Transformation: ML and Geospatial Techniques for Land Use/Land Cover Mapping in the Indian Sundarbans

Abstract The Indian Sundarban has the unique mangrove region and dynamic deltaic ecosystem zone that experience the ever-changing land use and land cover pattern (LULC) due to the environmental and man-made interactions. The LULC pattern assessment is very adequate for monitoring the biodiversity, ecological integrity, resource management in sustainable way as well as climate change impacts. The current chapters aimed to prepare the LULC map using multi-temporal satellite images through the Support Vector Machine (SVM) method to identify the change pattern of land use in Sundarban. The chapter has been identifying the change pattern of Indian Sundarban land use over 20 years by preparing the LULC high resolution map. The prepared LULC map accuracy has been checked by the Kappa Co-efficient method. The SVM classifier, known for its reliability to manage non-linear boundaries and very limited training data, exploring the high classification accuracy, validated using Kappa statistics and confusion matrix. Results demarked adequate spatiotemporal modifications, including degradation mangrove vegetation and expansion of settlements, and emphasize the need for sustainable coastal land management policies. This chapter exploring the potential of SVM-based classification for accurate LULC mapping in ecologically sensitive and susceptible coastal regions like the Sundarbans.

Keywords Landscape Transformation · Machin Learning · Geospatial Techniques · Land use/Land cover · Support Vector Machine (SVM) · Accuracy Assessment · Indian Sundarbans

6.1 Introduction

Ecological and changes is most vital issue to understand the changes on earth surface. The global environment rapidly altered based on the modification of land use and land cover that is a global concern (Jiang & Tian, 2010; Galata, 2020; Mallupattu & Reddy, 2013; Thakur et al., 2021). The world wise vegetation cover has been decreases due to the rapid growth of anthropogenic factors over last centuries (Thakur et al., 2021).

© The Author(s), under exclusive license to Springer Nature Switzerland AG 2026
A. Nath et al., *Shoreline Modification Over the Indian Sundarban*,
SpringerBriefs in GIS, https://doi.org/10.1007/978-3-032-13981-8_6

In case of coastal areas, physical and man-made both factors have been triggered for land use land cover changes especially in Mangrove vegetation areas natural processes are very vulnerable like soil salinity, coastal flooding, storm surges and water intrusion as natural factors (Khan et al., 2015; Das, 2022). In tropical and subtropical climate condition, the mangroves are mainly observed in the intertidal coastal zone. Near about 60–70% mangrove vegetation constitute the rich and biologically productive ecosystem (Clough, 1998). 7% Of South Asian mangrove has been covered in total world mangrove vegetation (Giri et al., 2015). Mangrove has a significant role to protect the coastal region from several natural hazard and disaster (Giri et al., 2007). Mangroves plays as significant assignment in South Asian countries to the ecosystem service i.e., stabilize the shoreline, reduce damages from storm and cyclones, increase the ground water quality and level, reduce the salt water intrusion, protect from flooding, decrease the microclimatic effect, balancing of global carbon requirement, nourishment of sediment transportation, reduce soil erosion and provide the recreation site and so on for man and environment (Blasco & Aizpuru, 2002; Duke et al., 2007; Giri et al., 2007; Pham et al., 2017). The Sundarbans mangrove forest has no exception to provide the all services to the environment and ecosystem (Giri et al., 2008; Mukul et al., 2020).

To adopt the environmental changes, mangroves are shifted towards sea or land. But the mangroves are found under immense threat of deforestation with the annual decrease of 1–2% caused by global climate change, natural disaster and human intervention (Sardar & Samadder, 2021). The largest mangrove forest has been observed over the Ganga, Brahmaputra and Meghna delta at India and Bangladesh respectively. The delta is significant for the large sedimentation carried from Himalayan region by these three rivers and developed widely (Datta & Deb, 2012; Ghosh et al., 2015; Islam, 2014). The entire mangrove also protects the natural resources and the ecosystem as well the coastal areas from several natural hazard and disaster (Giri et al., 2015; Mukul et al., 2020).

Sundarbans vegetation areas are mostly affected by the several disasters like coastal erosion, cyclone and storm surges and also various anthropogenic issues i.e., woodcutting, fishing, hunting and tourism. The mangrove density in Sundarbans has also reduced by the various diseases (Islam et al., 2019). To assess the loss and gain amount of land surface utilization, the land use land cover (LULC) is most significant tool that help to identify the characteristics and nature of land cover at any region. This assessment has also provided the insightful information about the planning of natural resources which helpful to the policy makers to utilize and management of the resources (Negassa et al., 2020; Mallupattu & Reddy, 2013, Rajendran et al., 2020; Periyasamy et al., 2021; Pan et al., 2017). Long term LULC modification has been associated with environmental changes and growth of urbanization as well the composition of land surface (Townshend et al., 1987). This LULC analysis also helpful for quantifying and observe the specific alteration of any region within any certain period (Perrigo et al., 2020; Scott et al., 2017). It is providing the pinpoint modification of land cover as well the change detection with large scale which helping what and where the changes made (Dao & Liou, 2015; Talukdar et al., 2020; Viana et al., 2019).

Spatio temporal land cover information has been collected from remote sensing and satellite imagery data (Bhattacharjee et al., 2021). In LULC assessment, machine learning algorithm plays a vital and significant role (Aryal et al., 2023). Support Vector Machine (SVM), Convolutional Neural Network (CNN) and Random Forest (RF), these are the most applied supervised classifier to analyse the land cover pattern (Jong et al., 2011). Precision, automation and scalability of machine learning provide the insightful information in environmental management, monitoring and urban planning in diverse areas for sustainable development (Ashraf et al., 2024; Rahman et al., 2010). The machine learning algorithm provide the precise accuracy through the understanding of complex pattern of satellite image with the addition of various mapping of vegetation, agricultural, urban and land surface patterns (Zafar et al., 2024a, 2024b). The machine learning algorithm can preserve the large excel datasets that efficient to analyse the change detection over certain time period (Stehman & Czaplewski, 1998). The versatility of these algorithms with accurate automation making the method more valuable for assessing and monitoring the environment, urban planning, policy making as well as sustainable development (Bechtel & Daneke, 2012).

To monitoring on the earth surface and capturing the information, remote sensing data plays a pivotal role through LULC analysis (Cihlar et al., 2003; Mehmood et al., 2022). Remotely sensed date like satellite imagery, aerial photos provide the insightful data for urban areas, vegetation cover, water areas and more (Lehnert et al., 2014). Temporal and spatial dataset can assess the changes over time period (Otukei & Blaschke, 2010). Land cover pattern and classes has been classified by supervised and unsupervised algorithms using remotely sensed data (Zafar, 2024). In mangrove forest, land cover modification is most crucial and significant due to its environmental impact and no exceptional for Indian Sundarbans. The spatial extension of mangrove has not been changing in Sundarbans in last few decades but the internal structure highly modified especially agricultural land to open mangrove forest area as well as the built-up areas (Giri et al., 2007). Mostly the modification observed in the buffer or transition zones of the forest where unsustainable livelihood has been enhanced in last few decades (Giri et al., 2007).

In the current past, various land use studies have been done on Sundarbans mangrove vegetation applying the remotely sensed dataset especially in Bangladesh part to assess the mangrove modification (Akhter, 2006; Diyan, 2011). From last few decades in Sundarbans, mangrove density, ecological disturbance, ecosystem condition, forest dynamic studies have been conducted over Sundarbans as well as physical and anthropogenic disturbances also studied (Mandal & Hosaka, 2020; Ishtiaque et al., 2016; Mukul et al., 2019; Dutta et al., 2015; Ghosh et al., 2016). Current research studies have been conducted on the mangrove forest with the spatial and temporal extension (Akhter, 2006; Diyan, 2011; Hasan et al., 2020; Islam et al., 2019; Salam et al., 2007). Few studies only focuses on limited part of Sundarbans (Kumar et al., 2021). But areal and temporal land dynamics studies are very limited in Sundarbans. The LULC research in Indian Sundarbans coastal areas are most significant applying the machine learning method with the advance algorithms that have accurate and precise classification of land pattern (Braun & Hochschild, 2017;

Bui & Mucsi, 2021; Liou et al., 2001; Rish, 2001; Roy et al., 2023). Support Vector Machine (SVM) has one of the most useful machine learning tool to acquire the complex land cover change patterns with enabling the automated extraction method from satellite data with high accuracy to understanding of any landscape change pattern (Boos & Kuang, 2010; Che et al., 2014; Hong et al., 2021; Lu & Weng, 2007; Natarajan et al., 2021).

The present chapter execute the land use land cover analysis through the SVM machine learning method using remotely sensed satellite imagery performing on the Geographical Information System (GIS) environment in Indian Sundarbans coastal part. To achieve the objective of the current research work, follows steps are (1) delineate the land cover pattern near coastal buffer zones from 2000 to 2020 that divided into ten years' interval (2000, 2010 and 2020). (2) Identify the modification of mangrove areas as well as built up. (3) Finally interpret the change analysis and comparing with different time period according to the outcome of the LULC mapping. This chapter is very significant for detect the modification over coastal part of Indian Sundarbans where shoreline change also very important impact on LULC alteration.

6.2 Applied Dataset and Processing

In present research work, Landsat-5 (LS5) and Landsat-8 OLI (LS8) satellite image has been used which obtained from USGS (United States Geological Survey) website (https://earthexplorer.usgs.gov). The study focuses on the LULC change between 2000 and 2020-time period. For these three different satellite images were acquired from USGS i.e., 2000 (Landsat-5), 2010 (Landsat-5) and 2020 (Landsat 8). All chosen data were cloud free for extract the land cover features. The accuracy assessment of LULC maps has been performed using ground truth data and Google earth image system. Through a common map projection system all data set has been layer stacking and rectified geometrically as WGS 1984 UTM zone 45N. The work follows the SVM machine learning process to execute the land use land cover mapping. To show the LULC changes six specific classes has been assessed for the Indian Sundarbans near coastal part where shoreline change is one of the major concerns. The entire land cover features are directly or indirectly associated with shoreline modification over the region. The chosen classes are healthy vegetation, vegetation, built up, sparsely vegetation, barren soil and water body. To execute the LULC process ArcGIS 10.8 platform has been used and training sample set were generated for each LULC map to identify the features. After choosing the training files, two formats have been used to save them i.e., particular signature file format and shapefile format. Then SVM algorithms has been applied to prepare the LULC maps using those signature files and output map has been saved to generate the final LULC classification map.

6.3 Methodology

The present assessment depicted by the four following major steps i.e., (1) integration of machine learning methods, (2) accuracy assessment of LULC maps by Kappa coefficient algorithms, (3) generate the LULC maps and (4) identify the spatial and temporal changes of each time period to another. The present study follows the SVM supervised machine learning methods to acquire the features from the satellite images (Kotsiantis et al., 2007).

6.3.1 Support Vector Machine

The SVM is a binary classification method that non-parametric machine learning supervised classification process. The SVM is founded by the idea of SRM (Structural risk minimization) that maximize and divides the data closest to the SAM (spectral angel mapper) of hyperplane's. This method implies by the division of data points into various land cover classes. The width of the margin has been maximized in the vectors while performing the SVM. Multiple variables that may continuous and categorical as well linear and non-linear can performed in SVM. SVM hyperplane or margins has been defined by the vector training samples. In case of remote sensing radial basis function (RBF) and polynomial kernels has been frequently applied. Comparing to other conventional method, RBF is most applied method for LULC classification which provides the maximum range of accuracy.

The SVM is determine to separate the hyperplane from the available hyperplanes. The actual SVM generated with the data set and identify the hyperplane that also divided the set of data into multiple classes. To reduce the errors and obtain the more precise map, the appropriate Kernel function has been required in SVM algorithm for generate the hyperplanes. It is a very crucial phenomenon in SVM method. The main goal of SVM is to find out the ideal boundary between all support vectors that will enhance the distance. SVM is machine learning supervised binary algorithm applied for regression task and LULC classification. The SVM follows the form binary classification formula:

$$\boldsymbol{f}(x) = \mathbf{Sin}\left(\sum_{i=1}^{n} \propto \boldsymbol{i y i K}(x, xi) + \boldsymbol{b}\right) \tag{6.1}$$

Here, f(x) represent the decision function which predicts the input classes, x, αi is the multipliers of language that acquired from training sample, yi is training sample of xi class label and K (x, xi) is Kernel function. The bias term is representing by b.

The sign function demarked the predicted class resting on the calculated decision function. Find out the optimal hyperplane which separate various classes in feature class space is the main aim of SVM.

6.3.2 Accuracy Assessment

Accuracy assessment is most significant for any remote sensing image classification. To determine the performance of algorithm with accuracy, precision and recall, the kappa coefficient is widely used accuracy assessment method (Mohamed et al., 2022). Kappa based accuracy analysis assess the overall accuracy and measure the validity of any algorithm of classification. To represent the precision or accuracy of land use mapping, the Kappa coefficient acquired by the confusion matrix. The accuracy assessment derived by the user accuracy, producer accuracy, overall accuracy and Kappa coefficient. True or exact value of a class has been determining by the producer accuracy, the false unfavorable values of that class has been demarked by the user accuracy, the entire map true favorable value has been calculated by overall accuracy. The Kappa coefficient has been calculated by the following formula (Das et al., 2021; Petropoulos et al., 2015; Verma et al., 2020):

$$\mathbf{K} = \frac{\sum \frac{u}{a} - 1\frac{C^a \sum \frac{U}{a} - 1\frac{C_a C_a}{Q^2}}{Q}}{1 - \sum \frac{U}{a} - 1\frac{C_{a.}C_a}{q2}} \tag{6.2}$$

Here, C_a = Row sum.

Overall accuracy of the LULC map has been calculated by the following equation:

$$\mathbf{OA} = \frac{\sum \frac{U}{a} - 1C_a}{Q} * 100\% \tag{6.3}$$

OA = Overall accuracy.

Here, total pixel and classes represented by Q and U respectively. The overall accuracy has been accepted 85% for any LULC mapping (Congalton & Green, 1999; Li & Zhou, 2009; Lu & Weng, 2007).

6.3.3 Detection of Land Transformation from 2000 and 2020

Transformation land refers to the modification of land by physical and functional characteristics. It means physical factor or ecosystem converted into built up, water or barren land, or modifying the way to cover the land. The land transformation has significant influence on the environment, economy and society could associate with mangrove loss, decline the biodiversity, soil degradation, climate change, ecosystem changes and habitat loss. To complete the layer-based examination of LULC from 2000–2020, ArcGIS toolbox has been used where a conversion tool employed to transform the LULC layers to vector from raster format. With the help reclassify option generated land use vector layers re-classified with the features from 2000–2020. By the assessment of LULC features separately, it can be possible to identify the certain changes or transformation of selected area from 2000–2020 land cover pattern.

6.4 Result of Analysis

A transformation scenario of land cover at any region can be possible to generate the land use land cover classification map. In the current research six land use category has been assessed for 20 years of time period (2000–2020) for the Indian Sundarbans mangrove area to determine the land cover changes. Three maps have been generated to identify the land transformation i.e., 2000, 2010 and 2020. To identify the prominent changes over the region two successive time frame has been chosen i.e., 2000–2010 and 2010–2020 duration 10 years and comparing the maps for change detection analysis. Overall change or transformation from 2000–2020 has also been calculated to show the overall changes within 20 years. The map of LULC 2000 has been shown in Fig. 6.1 and LULC map 2010 also shows in Fig. 6.2. The LULC map of 2020 has been shown in Fig. 6.3. After generating the map LULC transformation area has been calculated assess the change detection.

6.4.1 Land Cover Pattern of 2000–2010

Table 6.1 indicates the land cover pattern of Indian Sundarbans during 2000–2010-time period. From the analysis it can be seen that healthy vegetation area changed

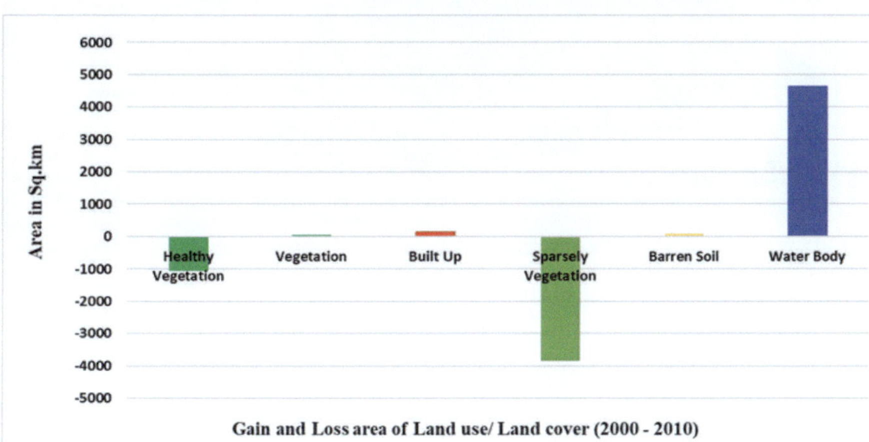

Fig. 6.1 Change statistics of LULC features from 2000 to 2010 (Loss and gain area)

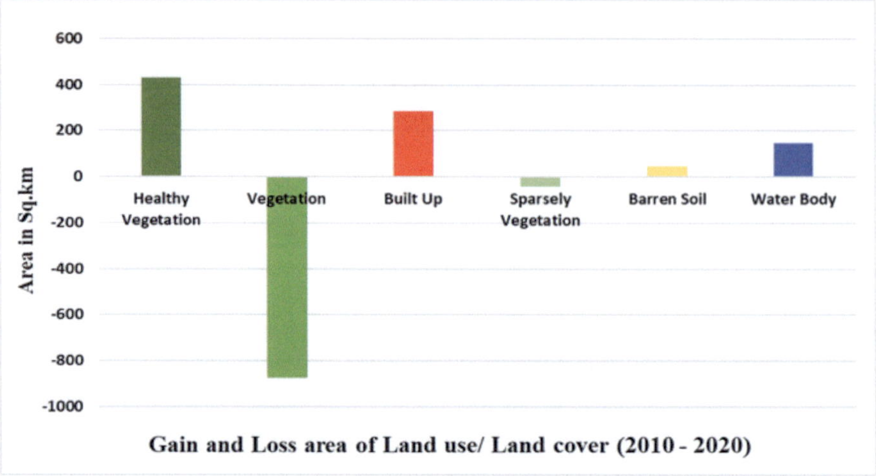

Fig. 6.2 Change statistics of LULC features from 2010 to 2020 (Loss and gain area)

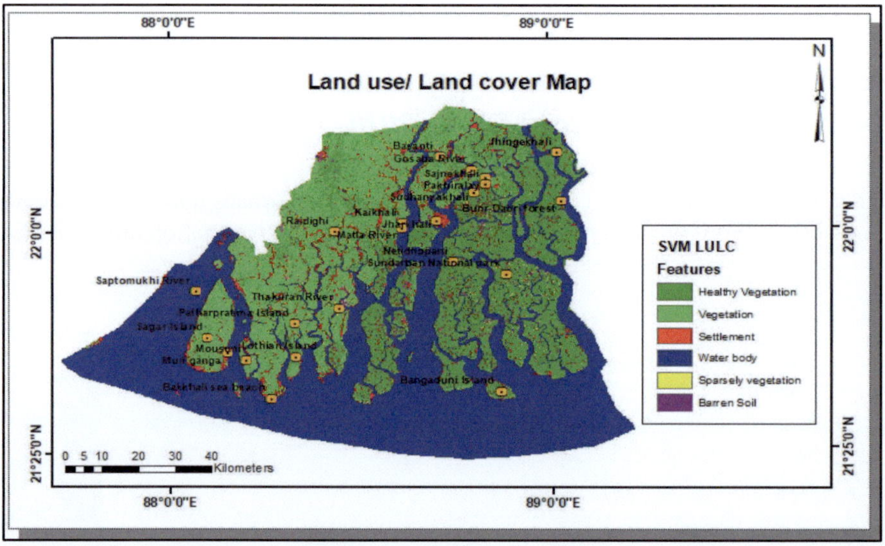

Fig. 6.3 Land use/land cover map of 2000

largely. From 2686.52 km^2 (27.40%) healthy vegetation (mangrove) area transformed to 1634.57 km^2 (16.67%) heathy vegetated region. This healthy vegetation area considered as core mangrove area in Sundarbans Vegetation. However, no such change is seen in the vegetation area over Indian Sundarbans where 1858.32 km^2 area of vegetation in 2000 converted as 1903.65 km^2 area in 2010. In built up area has also not much changed over 10 years, 401.55 km^2 area of built up expanded to

Table 6.1 Land use coverage area from 2000–2010 (km^2 and %)

Feature name	Area in km^2 2000	Area in % 2000	Area in km^2 2010	Area in % 2010	Net change area in km^2
Healthy vegetation	2686.52	27.40	1634.57	16.67	−1051.95
Vegetation	1858.32	18.95	1903.65	19.41	45.33
Built Up	401.55	4.09	528.77	5.39	127.21
Sparsely vegetation	4766.70	48.61	928.50	9.46	−3838.20
Barren soil	24.28	0.24	113.28	1.15	89.01
Water body	66.79	0.68	4696.40	47.89	4629.61

528.77 km^2 area. However, an adverse transformation is being observed in sparsely vegetated area during 2000–2010 where 4766.70 km^2 (48.61%) area reduced into 928.50 km^2 (9.46%) area of sparsely vegetation zone. The most significant change is demarked in the case of water bodies which are 66.79 km^2 in 2000 to 4696.40 km^2 in 2010. In case of barren soil 24.28 km^2 in 2000 area has been modified as 113.28 km^2 in 2010.

6.4.2 *Land Cover Pattern of 2010–2020*

Table 6.2 indicates that the healthy vegetated area has been increased slightly which are 16.67% in 2010 to 21.06% in 2020. But the graphical changes in vegetation area is being felt less where 1903.65 km^2 (19.41%) area in 2010 has been reduced as 1028.74 km^2 (10.49%) area in 2020. In terms of built up area, the area is changing from 528.77 km^2 (5.39%) in 2010 to 814.01 km^2 (8.30%) in 2020. Very little change is being observe in the case of water bodies, barren soil and sparsely vegetated area during 2010–2020 over Indian Sundarbans.

Table 6.2 Land use coverage area from 2010–2020 (km^2 and %)

Feature name	Area in km^2 2010	Area in % 2010	Area in km^2 2020	Area in % 2020	Net change area in km^2
Healthy vegetation	1634.57	16.67	2064.72	21.06	430.15
Vegetation	1903.65	19.41	1028.74	10.49	−874.91
Built Up	528.77	5.39	814.01	8.30	285.23
Sparsely vegetation	928.50	9.46	886.96	9.05	−41.54
Barren soil	113.28	1.15	161.37	1.64	48.09
Water body	4696.40	47.89	4843.67	49.42	147.26

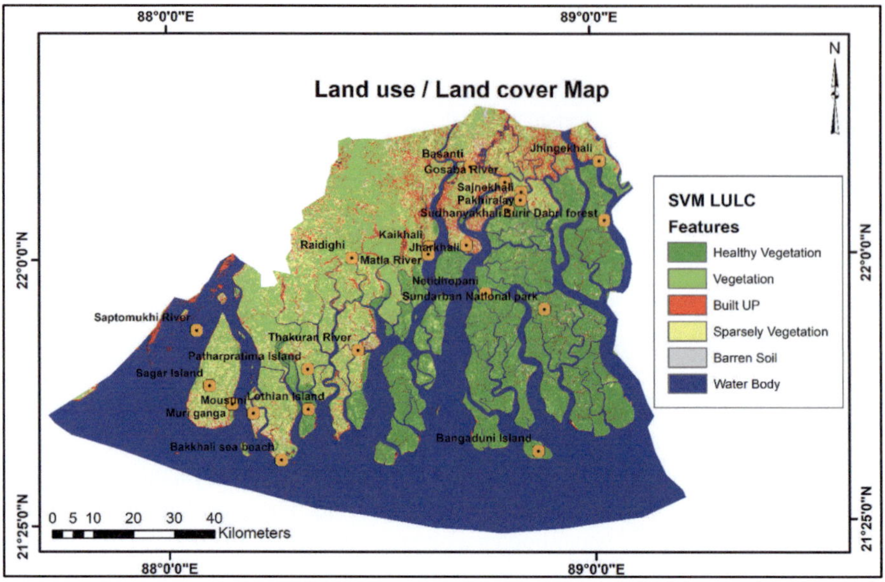

Fig. 6.4 Land use/land cover map of 2010

6.4.3 Change Detection of LULC from 2000–2010

Change detection statistics show how much area is changing after generating the each LULC map by SVM. Figure 6.1 demarked the loss and gain area of each features between 2000 and 2010-time span. Figure 6.1 gives an idea of change area statistics of all features within that successive 10 years' interval. Table 6.1 tabulates the LULC change scenario (2000–2010) and net change statistics of loss and gain area according to each feature in km^2. In this case, it appears that the healthy vegetation area (mangrove) has changed the most where -1051.95 km^2 area reduced in 2010 from 2000. A similar situation is observed in the sparsely vegetated area where $-$ 3838.20 km^2 area decreases between 10 years. But in the case of water bodies, a highly positive change is being observed, which is 4629.61 km^2. No such change was observed between built up, vegetation and barren soil area during these 10 years (2000–2010) (Table 6.2). The net change area in km^2 is shown in Fig. 6.1. Net area statistics has been calculated from Figs. 6.3 and 6.4 LULC map of Indian Sundarbans.

6.4.4 Change Detection of LULC from 2010–2020

Figure 6.2 denotes the net change of LULC features between 2010 and 2020-time span. The next successive 10 years (2010–2020) show the maximum changes in various feature areas. Firstly, very positive changes are being noticed in the healthy

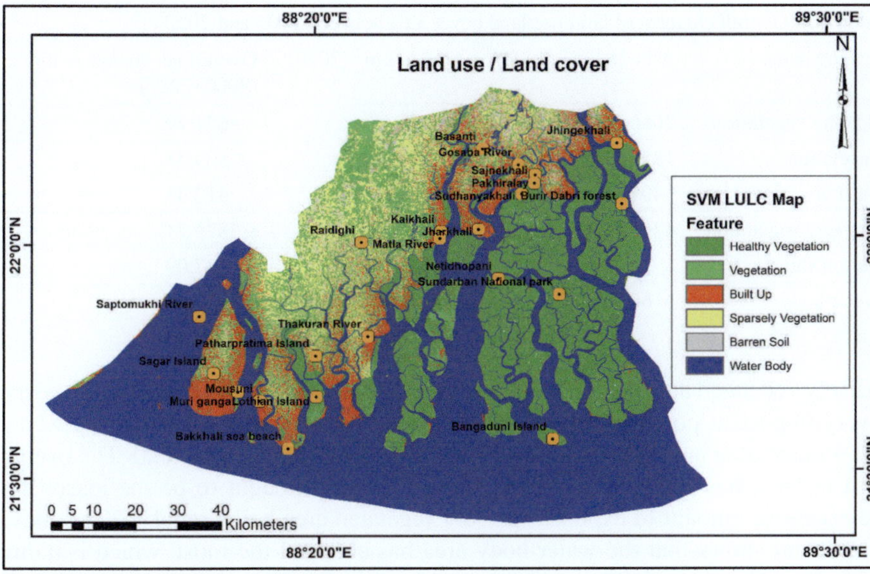

Fig. 6.5 Land use/land cover map of 2020

vegetated area between 2010 and 2020 where 430.15 km^2 area increased in these 10 years. But negative changes are clearly being observed in the area of vegetation and sparsely vegetated areas where −874.91 km^2 and −41.54 km^2 respectively reduced. But expansion of water body has not been exceeding much like past 10 years. Here, only 147.26 km^2 water body area has been expanded in these 10 years. The net change area has been calculated from the base map of LULC (Figs. 6.4 and 6.5) through SVM.

6.4.5 *Overall Net Changes of LULC from 2000–2020*

The overall net change analysis between 2000 and 2020 will show how much land cover features change has occurred in any part of Indian Sundarbans. Above all, Table 6.3 shows the overall land use/land cover net change or loss and gain area between 2000 and 2020. In this case, it can be seen that healthy vegetation was 2686.52 km^2 in 2000 which has changed to 2064.72 km^2 in 2020. So, the overall net change in healthy vegetation is going to be 621.79 km^2 that indicate a prominent decrease of core vegetated area (mangroves). A similar situation is seen in the case of vegetation and sparsely vegetated areas. 1858.32 km^2 vegetated area in 2000 has been reduce to 1028.74 km^2 vegetated area in 2020. Here, −829.58 km^2 area has lost within 20 years. 4766.70 km^2 sparsely vegetated area in 2000 has been decreased in 2020 as 886.96 km^2. Here, a very significant negative change has observed in

Table 6.3 Overall changes of land use/land cover area between 2000 and 2020

Feature name	Area in km^2 (2000)	Area in km^2 (2020)	Overall net change in km^2 (2000–2020)
Healthy vegetation	2686.52	2064.72	−621.79
Vegetation	1858.32	1028.74	−829.58
Built up	401.55	814.01	412.45
Sparsely vegetation	4766.70	886.96	−3879.74
Barren soil	24.28	161.37	137.09
Water body	66.79	4843.67	4776.88

sparsely vegetated area that −3879.74 km^2. But during the aforementioned 20 years, a very important positive change is being observed in the case of built up where 401.55 km^2 area in 2000 positively shifted to 814.01 km^2 area in 2020. The overall built up area has changed as 412.45 km^2. Which is thought to be the reason for the excessive amount of exposed sparsely vegetated area being reduced. The above assessment shows that the water body area has changed the most, which is a dire indication. The area of the water body in 2000 was 66.79 km^2, which has changed to 4843.67 km^2. Overall 4776.88 km^2 area has been induced in water bodies that is the triggering sign for near future in Indian Sundarbans. Figure 6.6 represents the occupied area of land cover features between 2000 and 2020 and how much it has changed.

6.4.6 *Result of Accuracy Assessment*

No LULC classified map is complete until its accuracy assessment is calculated (Bradley, 2009). Accuracy assessment is a very important method to verify the authenticity of all the information in the above LULC map. The land cover information acceptability is determined by the producer's accuracy, user's accuracy and overall accuracy. From the analysis it has been remarked that healthy vegetation, vegetation and water body acquired the maximum producer's accuracy in 2000, 2010 and 2018. Some error found in built up areas where 85% accuracy observed in case of 2020 but in 2000 and 2010 found maximum producer's accuracy. In case of user's accuracy, maximum accuracy has been observed in healthy vegetation, vegetation and sparsely vegetated area. A complete analysis shows that overall accuracy was 85% for 2000, 95% in 2010 and 90% in 2020. Kappa coefficient analysis shows that 85.25% of the accuracy obtained in 2000, 93.67% accuracy in 2010 and 92.47% accuracy determined in 2020 (Table. 6.4).

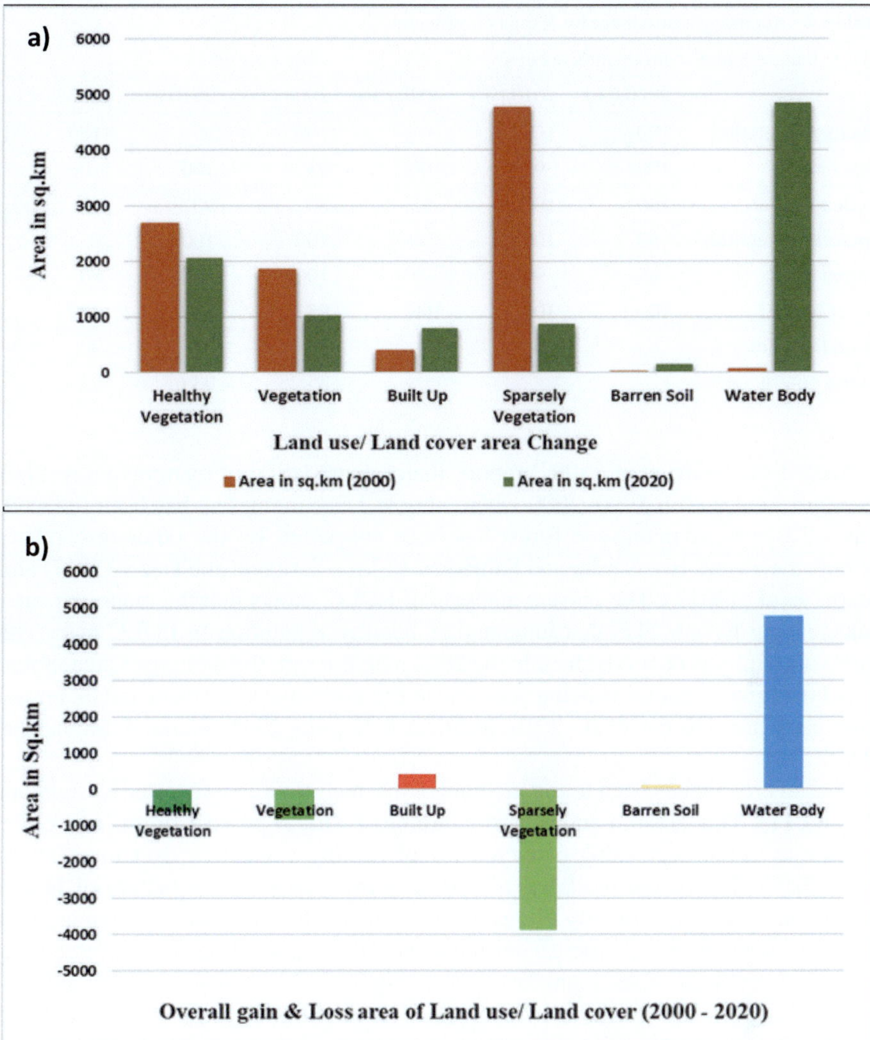

Fig. 6.6 Land use/land cover change (**A**) & Overall loss and gain area (**B**) between 2000 and 2020

6.5 Discussion

A change is being observed in every feature of the Sundarbans mangrove. Each land cover category has changed to another land cover category over time. But the core mangrove (healthy vegetation) area has changed very few portions, only some changes are observed near the river line and coastline areas. But most of the land cover changes are occurring the buffer area of mangroves, due to the human intervention and natural hazard.

Table 6.4 Accuracy assessment by Kappa co-efficient

LULC feature name	Producer's accuracy			User's accuracy		
	2000 (%)	2010 (%)	2020 (%)	2000 (%)	2010 (%)	2020 (%)
Healthy vegetation	100	100	90	100	95	100
Vegetation	100	90	100	95	100	100
Built up	100	100	85	80	90	85
Sparsely vegetation	85	100	90	100	100	95
Barren soil	85	90	95	100	100	80
Water body	100	100	100	95	88	100
Overall accuracy				85	95	90
Kappa co-efficient				85.25	93.67	92.47

According to Giri et al. (2007) report, major degradation of mangrove forest has been indicated within 1990–2000 where almost 1.2% mangrove has been reduced. This 1.2% loss of mangrove forest has been supported by the other researchers through their work like Sardar and Samaddar (2021), Samanta and Hazra (2017) and Quader et al., (2017). The current analysis of LULC shows that the mangrove area is decreasing by −6.34% that indicated as healthy vegetation in LULC map. The excess alkalinity is observed here in the 2020 map through the mangrove area. Some loss of mangrove species is being seen due to the excessive settlement and increased water body area (Karim et al., 2019; Rahman & Begum, 2011; Sardar & Samadder, 2021).

Coastal erosion has been found in Sundarbans mangrove area reported by Rahman et al. (2011). This erosion and accretion are most natural process of any mangrove region due to the decreasing or increasing of healthy vegetation over time (Islam et al., 2019). The recent work has also been shown the land use pattern of Indian Sundarbans every prominently. Here too, it is seen that the intrusion of major amount of alkaline water gradually increasing and most of the water volume is seen increasing in the deepest areas of the mangrove forest (Fig. 6.5). Most alkaline water appears to increasing on the northern side of study area. Similarly, the core mangrove area (healthy vegetation) is expected to increase over the next 2010–2020 due to various sedimentation activities (Table. 6.2). Some previous studies have shown that mangroves have been increased to some extent by this time period (Islam et al., 2019). On the other hand, another study shows that every year, 10% of the mangrove forest of the Sundarbans is partially destroyed due to various types of cyclonic storms (Mandal & Hosaka, 2020). But the river, through its function and through the process of sedimentation, helps to regenerate the forest land in Sundarbans. Kumar et al. (2021), study looked at how mangrove vegetation is decreasing and water intrusion gradually increasing where 0.44% vegetation lost within 1989–2019 and shifted to water body of 1.3%. the current research work also revealed how vegetation cover areas are being reduced and turned into water bodies where −8.46% vegetated area has been lost between 20 years. The vegetation covers in 2020 was severely impacted by cyclonic Amphan, which directly hit the Sundarbans mangroves at 20th may 2020. A map from December 2020 was used for the current research to better understand the damage of mangrove forest. It is clearly seen that the vegetation covers in 2010 has decreased significantly from what it was in 2020 and this decrease is −874.91 km^2 of area.

Another study showed that saline water intrusion is one of the main reasons for reducing the mangrove density in Sundarbans (Quader et al., 2017) and cyclonic storms is behind on water intrusion. The extreme salt water intrusion creates a barrier between the fresh water and sea water, which directly affects the health of the Sundarbans mangrove. In the current study, the increase in the built-up area along with the increase in water volume is also being gradually observed, which is becoming one of the reasons for the damage of the Sundarbans, here 4.21% built up increased within 20 years.

The Sundarbans is a place where national and international tourism are very active and tourist attract to visit. This additional amount of tourism is increasing, with tourists coming here and the mangroves are becoming an alarming situation due to the overcrowding compared to their capacity (Amin, 2018; Begum, 2015). Tourism is one of the major reasons for river bank erosion. Through all the above discussion and analysis, it is seen that the increase of water content, enhancement of built up area and decrease in vegetation area of the Indian Sundarbans is also becoming a major concern. Mangrove forests are turning into non-vegetated and water bodies, which is a matter of great concern.

6.6 Conclusion

The current research work attempts to assess the land cover status of the Indian Sundarbans between 2000 and 2020 using Support Vector Machine (SVM) method of machine learning approach through three individual satellite imagery. The recent study shows that some damage is occurring to mangrove forests here due to the various human interventions. Yet it appears that core part of the mangrove remains unchanged. The present study reveals that the normal vegetated area of the south part of Indian Sundarbans has suffered some damage, but the damage to the core area of mangrove (indicated by healthy vegetation in LULC map) area is less. The mangrove area has mainly damaged near the river bank and estuary areas where cyclonic attacks trigger a large amount of erosion every year. Just as the amount of erosion seen here is due to cyclonic action, the land can also regenerate here as the result of sedimentation (Bomer et al., 2020). The rivers in the Sundarbans region carry a large amount of sediment, which helps to flourish the mangrove trees. The analysis shows that the changes seen in the core mangrove area are only due to some land damage caused by excessive saline water intrusion through the riverbed. But one of the causes of vegetation loss is the excessive amount of built up accumulation. Through the above analysis, a prominent idea can be made about the condition of the mangroves in the Indian Sundarbans region and the damage caused to them.

References

Akhter, M. (2006). Remote sensing for developing an operational monitoring scheme for the Sundarban Reserved Forest, Bangladesh (dissertation). Germany (Dresdren): Technische Universität Dresden.

Amin, M. R. (2018). Sustainable tourism development in Sundarbans, Bangladesh (a world heritage site): Issues and actions. *Journal of Business Studies, 39*(2), 31–52.

Aryal, J., Sitaula, C., & Frery, A. C. (2023). Land use and land cover (LULC) performance modeling using machine learning algorithms: A case study of the city of Melbourne, Australia. *Scientific Reports, 13*(1), 13510.

Ashraf, T., Aslam, J., Mehmood, M. S., Ahamad, M. I., & Rehman, A. (2024). Geospatial assessment of built environment on land surface temperature in district Sheikhupura, Punjab, Pakistan. *Discover Geoscience, 2*(1), 30.

Bechtel, B., & Daneke, C. (2012). Classification of local climate zones based on multiple earth observation data. *IEEE Journal of Selected Topics in Applied Earth Observations and Remote Sensing, 5*(4), 1191–1202.

Begum, S. (2015). *Handbook on carrying capacity & management on tourist spots of the Sundarbans*. Khulna: Centre for Integrated Studies on the Sundarbans (CISS).

Bhattacharjee, S., Islam, M. T., Kabir, M. E., & Kabir, M. M. (2021). Land-use and land-cover change detection in a North-Eastern Wetland Ecosystem of Bangladesh using remote sensing and GIS techniques. *Earth Systems and Environment, 5*(2021), 319–340. https://doi.org/10.1007/s41748-021-00228-3.

Blasco, F., & Aizpuru, M. (2002). Mangroves along the coastal stretch of the Bay of Bengal: Present status. *Indian Journal of Marine Sciences, 31*(1), 9–20.

Bomer, E., Wilson, C., Hale, R., Hossain, A., & Rahman, F. (2020). Surface elevation and sedimentation dynamics in the Ganges-Brahmaputra tidal delta plain, Bangladesh: Evidence for mangrove adaptation to human-induced tidal amplification. *CATENA, 187*, Article 104312. https://doi.org/10.1016/j.catena.2019.104312.

Boos, W. R., & Kuang, Z. (2010). Dominant control of the South Asian monsoon by orographic insulation versus plateau heating. *Nature, 463*(7278), 218–222.

Bradley, B. A. (2009). Accuracy assessments of mixed land cover using a GIS-designed sampling scheme. *International Journal of Remote Sensing, 30*(13), 3515–3529. https://doi.org/10.1080/01431160802562263.

Braun, A., & Hochschild, V. (2017). A SAR-based index for landscape changes in African savannas. *Remote Sensing, 9*(4), 359.

Bui, D. H., & Mucsi, L. (2021). From land cover map to land use map: A combined pixel-based and object-based approach using multi-temporal Landsat data, a random forest classifier, and decision rules. *Remote Sensing, 13*(9), 1700.

Che, T., Xiao, L., & Liou, Y. A. (2014). Changes in glaciers and glacial lakes and the identification of dangerous glacial lakes in the Pumqu River Basin, Xizang (Tibet). *Advances in Meteorology, 2014*(1), Article 903709.

Cihlar, J., Latifovic, R., Beaubien, J., Guindon, B., & Palmer, M. (2003). Thematic Mapper (TM) based accuracy assessment of a land cover product for Canada derived from spot vegetation (VGT) data. *Canadian Journal of Remote Sensing, 29*(2), 154–170.

Clough, B. (1998). Mangrove forest productivity and biomass accumulation in Hinchinbrook Channel, Australia. *Mangroves Salt Marshes, 2*, 191–198. https://doi.org/10.1023/A:1009979610871.

Congalton, R. G., & Green, K. (1999). *Assessing the accuracy of remotely sensed data: Principles and practices*. Boca Raton, Florida: CRC Press, Taylor & Francis Group.

Dao, P. D., & Liou, Y. A. (2015). Object-based flood mapping and affected rice field estimation with Landsat 8 OLI and MODIS data. *Remote Sensing, 7*(5), 5077–5097.

Das, G.K. (2022). Climate change and coastal hazards in Sunderbans. In: *Coastal Environments of India*. Springer Water. Springer, Cham. https://doi.org/10.1007/978-3-031-18846-6_9.

Das, N., Mondal, P., Sutradhar, S., & Ghosh, R. (2021). Assessment of variation of land use/land cover and its impact on land surface temperature of Asansol subdivision. *The Egyptian Journal of Remote Sensing and Space Science, 24*(1), 131–149. https://doi.org/10.1016/j.ejrs.2020.05.001.

Datta, D., & Deb, S. (2012). Analysis of coastal land use/land cover changes in the Indian Sundarbans using remotely sensed data. *Geo-Spatial Information Science, 15*(4), 241–250. https://doi.org/10.1080/10095020.2012.714104.

De Jong, R., de Bruin, S., de Wit, A., Schaepman, M. E., & Dent, D. L. (2011). Analysis of monotonic greening and browning trends from global NDVI time-series. *Remote Sensing of Environment, 115*(2), 692–702.

Diyan, M. A. A. (2011). Multi-scale vegetation classification using earth observation data of the Sundarban mangrove forest, Bangladesh. M.Sc. thesis, Universidade nova de lisboa, Portugal (Lisbon).

Duke, N. C., Meynecke, J. O., Dittmann, S., Ellison, A. M., Anger, K., Berger, U., et al. (2007). A world without mangroves? *Science, 317*(5834), 41–42. https://doi.org/10.1126/science.317.5834.41b.

Dutta, D., Das, P. K., Paul, S., Sharma, J. R., & Dadhwal, V. K. (2015). Assessment of ecological disturbance in the mangrove forest of Sundarbans caused by cyclones using MODIS time-series data (2001–2011). *Natural Hazards, 79*(2), 775–790.

Galata, A. (2020). Analysis of land use/land cover changes and their causes using landsat data in hangar watershed, Abay basin, Ethiopia. *Journal of Sedimentary Environments, 5*(4), 415–423. https://doi.org/10.1007/s43217-020-00025-4.

Ghosh, A., Schmidt, S., Fickert, T., & Nüsser, M. (2015). The Indian Sundarban mangrove forests: History, utilization, conservation strategies and local perception. *Diversity, 7*(2), 149–169. https://doi.org/10.3390/d7020149.

Ghosh, M. K., Kumar, L., & Roy, C. (2016). Mapping long-term changes in mangrove species composition and distribution in the Sundarbans. *Forests, 7*(12), 305.

Giri, C., Pengra, B., Zhu, Z., Singh, A., & Tieszen, L. (2007). Monitoring mangrove forest dynamics of the Sundarbans in Bangladesh and India using multi-temporal satellite data from 1973 to 2000. *Estuarine, Coastal and Shelf Science, 73*(1–2), 91–100. https://doi.org/10.1016/j.ecss.2006.12.019.

Giri, C., Zhu, Z., Tieszen, L., Singh, A., Gillette, S., & Kelmelis, J. (2008). Mangrove forest distributions and dynamics (1975–2005) of the tsunami-affected region of Asia. *Journal of Biogeography, 35*(3), 519–528. https://doi.org/10.1111/j.1365-2699.2007.01806.x.

Giri, C., Long, J., Abbas, S., Murali, R., Qamer, F., Pengra, B., & Thau, D. (2015). Distribution and dynamics of mangrove forests of South Asia. *Journal of Environmental Management, 148*, 101–111. https://doi.org/10.1016/j.jenvman.2014.01.020.

Hasan, M. E., Nath, B., Sarker, A. H. M., Wang, Z., Zhang, L., Yang, X., et al. (2020). Applying multi-temporal Landsat satellite data and Markov-cellular automata to predict forest cover change and forest degradation of Sundarban reserve forest, Bangladesh. *Forests, 11*(9), 1016. https://doi.org/10.3390/f11091016.

Hong, D., Hu, J., Yao, J., Chanussot, J., & Zhu, X. X. (2021). Multimodal remote sensing benchmark datasets for land cover classification with a shared and specific feature learning model. *International Social Photogrammetry, 178*, 68–80.

Ishtiaque, A., Myint, S. W., & Wang, C. (2016). Examining the ecosystem health and sustainability of the world's largest mangrove forest using multi-temporal MODIS products. *Science of the Total Environment, 569*, 1241–1254.

Islam, M. T. (2014). Vegetation changes of Sundarbans based on landsat imagery analysis between 1975 and 2006. *Landscape & Environment, 8*(1), 1–9.

Islam, M. M., Borgqvist, H., & Kumar, L. (2019). Monitoring mangrove forest landcover changes in the coastline of Bangladesh from 1976 to 2015. *Geocarto International, 34*(13), 1458–1476. https://doi.org/10.1080/10106049.2018.1489423.

Jiang, J., & Tian, G. (2010). Analysis of the impact of land use/land cover change on land surface temperature with remote sensing. *Procedia Environmental Sciences, 2*, 571–575. https://doi.org/10.1016/j.proenv.2010.10.062.

Karim, M. F., Zhang, X., & Li, R. (2019). Dynamics of shrimp farming in the Southwestern Coastal districts of Bangladesh using a shrimp yield dataset (SYD) and landsat satellite archives. *Sustainability, 11*(17), 4635. https://doi.org/10.3390/su11174635.

Khan, M. M. H., Bryceson, I., Kolivras, K. N. (2015). Natural disasters and land-use/land-cover change in the southwest coastal areas of Bangladesh. *Regional environmental change, 15*, 241–250. https://doi.org/10.1007/s10113-014-0642-8.

Kotsiantis, S. B., Zaharakis, I., & Pintelas, P. (2007). Supervised machine learning: A review of classification techniques. *Emerging Artificial Intelligence Applications in Computer Engineering, 160*(1), 3–24.

Kumar, M., Mondal, I., & Pham, Q. B. (2021). Monitoring forest landcover changes in the Eastern Sundarban of Bangladesh from 1989 to 2019. *Acta Geophysica, 69*, 561–577. https://doi.org/10.1007/s11600-021-00551-3.

Lehnert, L. W., Meyer, H., Meyer, N., Reudenbach, C., & Bendix, J. (2014). A hyperspectral indicator system for rangeland degradation on the Tibetan Plateau: A case study towards spaceborne monitoring. *Ecological Indicators, 39*, 54–64.

Li, B., & Zhou, Q. (2009). Accuracy assessment on multi-temporal land-cover change detection using a trajectory error matrix. *International Journal of Remote Sensing, 30*(5), 1283–1296. https://doi.org/10.1080/01431160802474022.

Liou, Y. A., Liu, S. F., & Wang, W. J. (2001). Retrieving soil moisture from simulated brightness temperatures by a neural network. *IEEE Transactions on Geoscience and Remote Sensing, 39*(8), 1662–1672.

Lu, D., & Weng, Q. (2007). A survey of image classification methods and techniques for improving classification performance. *International Journal of Remote Sensing, 28*(5), 823–870. https://doi.org/10.1080/01431160600746456.

Mallupattu, P. K., & Reddy, J. R. S. (2013). Analysis of land use/land cover changes using remote sensing data and GIS at an urban area, Tirupati, India. *The Scientific World Journal, 2013*, Article 268623. https://doi.org/10.1155/2013/268623.

Mandal, M. S. H., & Hosaka, T. (2020). Assessing cyclone disturbances (1988–2016) in the Sundarbans mangrove forests using Landsat and Google Earth Engine. *Natural Hazards, 102*(1), 133–150.

Mehmood, M. S., Zafar, Z., Sajjad, M., Hussain, S., Zhai, S., & Qin, Y. (2022). Time series analyses and forecasting of surface urban Heat Island intensity using Arima model in Punjab, Pakistan. *Land, 12*(1), 142.

Mohamed, M., Morsy, S., & El-Shazly, A. (2022). Improvement of 3D LiDAR point cloud classification of urban road environment based on random forest classifier. *Geocarto International*, 1–23. https://doi.org/10.1080/10106049.2022.2102218.

Mukul, S. A., Alamgir, M., Sohel, M. S. I., Pert, P. L., Herbohn, J., Turton, S. M., et al. (2019). Combined effects of climate change and sea-level rise project dramatic habitat loss of the globally endangered Bengal tiger in the Bangladesh Sundarbans. *Science of the Total Environment, 663*, 830–840. https://doi.org/10.1016/j.scitotenv.2019.01.383.

Mukul, S. A., Huq, S., Herbohn, S. A., Seddon, N., & Laurance, W. F. (2020). Saving the Sundarbans from development. *Science, 368*(6496), 1198–1198. https://doi.org/10.1126/science.abb9448.

Natarajan, L., Sivagnanam, N., Usha, T., Chokkalingam, L., Sundar, S., Gowrappan, M., & Roy, P. D. (2021). Shoreline changes over last five decades and predictions for 2030 and 2040: A case study from Cuddalore, southeast coast of India. *Earth Science Informatics, 14*, 1315–1325.

Negassa, M. D., Mallie, D. T., & Gemeda, D. O. (2020). Forest cover change detection using Geographic Information Systems and remote sensing techniques: A spatio-temporal study on Komto Protected forest priority area, East Wollega Zone, Ethiopia. *Environmental Systems Research, 9*(1), 1. https://doi.org/10.1186/s40068-020-0163-z.

Otukei, J. R., & Blaschke, T. (2010). Land cover change assessment using decision trees, support vector machines and maximum likelihood classification algorithms. *International Journal of Applied Earth Observation, 12*, S27–S31.

Pan, T., Zou, X., Liu, Y., Wu, S., & He, G. (2017). Contributions of climatic and non-climatic drivers to grassland variations on the Tibetan Plateau. *Ecological Engineering, 108*, 307–317.

Periyasamy, R., Roy, P. D., Chokkalingam, L., Natarajan, L., Sundar, S., Moorthy, P., & Gowrappan, M. (2021). Transformation analysis on landuse/land cover changes for two decades between 1999 and 2019 CE with reference to aquaculture–Nagapattinam Coast, Southeast India. *Journal of the Indian Society of Remote Sensing, 49*, 2831–2845.

Perrigo, A., Hoorn, C., & Antonelli, A. (2020). Why mountains matter for biodiversity. *Journal of Biogeography, 47*(2), 315–325.

Petropoulos, G., Kalivas, D., Georgopoulou, I. A., & Srivastava, P. (2015). Urban vegetation cover extraction from hyperspectral imagery and geographic information system spatial analysis techniques: Case of Athens, Greece. *Journal of Applied Remote Sensing, 9*, 096088–096105. https://doi.org/10.1117/1.JRS.9.096088.

Pham, V., Luu, V., Nguyen, T., & Koji, O. (2017). Will restored mangrove forests enhance sediment organic carbon and ecosystem carbon storage? *Regional Studies in Marine Science, 14*, 43–52. https://doi.org/10.1016/j.rsma.2017.05.003.

Quader, M. A., Agrawal, S., & Kervyn, M. (2017). Multi-decadal land cover evolution in the Sundarban, the largest mangrove forest in the world. *Ocean and Coastal Management, 139*, 113–124. https://doi.org/10.1016/j.ocecoaman.2017.02.008.

Rahman, M., & Begum, S. (2011). Land cover change analysis around the Sundarbans mangrove forest of Bangladesh using remote sensing and GIS application. *Journal of Science Foundation, 9*(1–2), 95–107. https://doi.org/10.3329/jsf.v9i1-2.14652.

Rahman, A. F., Dragoni, D., & El-Marri, B. (2011). Response of the Sundarban's coastline to sea level rise and decreased sediment flow: A remote sensing assessment. *Remote Sensing of Environment, 115*, 3121–3128. https://doi.org/10.1016/j.rse.2011.06.019.

Rahman, A., Aggarwal, S. P., Netzband, M., & Fazal, S. (2010). Monitoring urban sprawl using remote sensing and GIS techniques of a fast-growing urban centre India. *IEEE Journal of Selected Topics in Applied Earth Observations and Remote Sensing, 4*(1), 56–64.

Rajendran, G. B., Kumarasamy, U. M., Zarro, C., Divakarachari, P. B., & Ullo, S. L. (2020). Land-use and land-cover classification using a human group-based particle swarm optimization algorithm with an LSTM classifier on hybrid pre-processing remote-sensing images. *Remote Sensing, 12*(24), 4135.

Rish, I. (2001). An empirical study of the naive Bayes classifier. *IJCAI 2001 Workshop on Empirical Methods in Artificial Intelligence, 3*(22), 41–46.

Roy, P. D., Natarajan, L., Garcia-Arriola, O. A., & Chokkalingam, L. (2023). Anthropogenic impact in and around Lake Coatetelco (south-central Mexico) from land use and land cover analysis between 1980 and 2020 CE. *Journal of South American Earth Sciences, 129*, Article 104495.

Salam, M. A., Ross, L. G., & Beveridge, C. M. C. (2007). The use of GIS and remote sensing techniques to classify the Sundarbans Mangrove vegetation. *Journal of Agroforestry and Environment, 1*(1), 7–15. ISSN: 1995-6983.

Samanta, K., & Hazra, S. (2017). Mangrove forest cover changes in Indian Sundarban (1986–2012) using remote sensing and GIS. In: S. Hazra, A. Mukhopadhyay, A. Ghosh, D. Mitra, V. Dadhwal (Eds.), *Environment and earth observation. Springer remote sensing/photogrammetry*. Cham: Springer. https://doi.org/10.1007/978-3-319-46010-9_7.

Sardar, P., & Samadder, S. (2021). Understanding the dynamics of landscape of greater Sundarban area using multi-layer perceptron Markov chain and landscape statistics approach. *Ecological Indicators, 121*, Article 106914. https://doi.org/10.1016/j.ecolind.2020.106914.

Scott, G. J., England, M. R., Starms, W. A., Marcum, R. A., & Davis, C. H. (2017). Training deep convolutional neural networks for land–cover classification of high-resolution imagery. *IEEE Geoscience and Remote Sensing Letters, 14*(4), 549–553.

Stehman, S. V., & Czaplewski, R. L. (1998). Design and analysis for thematic map accuracy assessment: Fundamental principles. *Remote Sensing of Environment, 64*(3), 331–344.

Talukdar, S., Singha, P., Mahato, S., Pal, S., Liou, Y. A., & Rahman, A. (2020). Land-use land-cover classification by machine learning classifiers for satellite observations–A review. *Remote Sensing, 12*(7), 1135.

Thakur, S., Maity, D., Mondal, I., Basumatary, G., Ghosh, P. B., Das, P., & De, T. K. (2021). Assessment of changes in land use, land cover, and land surface temperature in the mangrove forest of Sundarbans, northeast coast of India. *Environment, Development and Sustainability, 23*, 1917–1943. https://doi.org/10.1007/s10668-020-00656-7.

Townshend, J. R., Justice, C. O., & Kalb, V. (1987). Characterization and classification of South American land cover types using satellite data. *International Journal of Remote Sensing, 8*(8), 1189–1207.

Verma, P., Raghubanshi, A., Srivastava, P. K., & Raghubasni, A. S. (2020). Appraisal of kappa-based metrics and disagreement indices of accuracy assessment for parametric and nonparametric techniques used in LULC classification and change detection. *Modeling Earth Systems and Environment, 6*, 1045–1059. https://doi.org/10.1007/s40808-020-00740-x.

Viana, C. M., Girão, I., & Rocha, J. (2019). Long-term satellite image time-series for land use/land cover change detection using refined open-source data in a rural region. *Remote Sensing, 11*(9), 1104.

Zafar, Z. (2024). Assessment of urbanization impacts on vegetation cover in major cities of Pakistan: Evidence from remotely sensed data. *GeoJournal, 89*(4), 1–21.

Zafar, Z., Zubair, M., Zha, Y., Fahd, S., & Nadeem, A. A. (2024a). Performance assessment of machine learning algorithms for mapping of land use/land cover using remote sensing data. *The Egyptian Journal of Remote Sensing and Space Sciences, 27*(2), 216–226.

Zafar, Z., Zubair. M., Zha, Y., Mehmood, M. S., Rehman, A., Fahd, S., & Nadeem, A. A. (2024b). Predictive modeling of regional carbon storage dynamics in response to land use/land cover changes: An InVEST-based analysis. *Ecological Informatics*, 102701.

Chapter 7
Exploring the Potential of Erosion-Accretion Zonation Mapping to Assess Short and Long-Term Shoreline Changes

Abstract To manage and understand the shoreline's versatility in highly sensitive and dynamic morphological coastal regions like Sundarban, erosion-accretion zonation mapping is the most critical and strategic tool. Short-term shoreline changes zonation mapping recorded the urgent geomorphic modification driven by storm surges, cyclones, saltwater intrusion, flooding, tidal fluctuation, ecosystem change, embankment reinforcement, and livelihood rescue. On the other hand, long-term zonation revealed the erosion-accretion trend, insightful planning of sustainable land use patterns, biodiversity restoration, climate change adaptation, and mangrove regeneration. With the combination of short and long-term zonation mapping, temporary to permanent trends of shoreline modification have been analyzed and demarked the hotspot erosional zones in the Indian Sundarbans region. From the zonation strategic tool, sediment accumulation areas have also been reached quickly, so the mangrove can easily regenerate. Integrating both long and short-term zonation mapping increases resilience by aligning urgent response strategies with maximum sustainability goals in this climate-susceptible deltaic mangrove ecosystem.

Keywords Erosion-accretion Zonation Mapping · Shoreline Shifting · Short-term Shoreline Changes · Long-term Zonation · Sustainability · Deltaic Mangrove Ecosystem · Indian Sundarban

7.1 Introduction

The coastal region is demarcated as the most significant area for human development and socio-economic development; hence, this region is under vulnerable conditions globally (Nichollos et al., 2007). The coastal region is highly susceptible in nature and also very dynamic (Kantamaneni et al., 2018; Lins-de-Barros, 2017; Masselink & Hughes, 2014). In coastal areas, hazards and disasters are very common triggering factors increasing day by day (Arkema et al., 2013; Donchyts et al., 2016; Shepard et al., 2011). In the current context, coastal erosion is manifest as a very serious hazard condition (Leatherman et al., 2000; Masselink & Russell,

© The Author(s), under exclusive license to Springer Nature Switzerland AG 2026
A. Nath et al., *Shoreline Modification Over the Indian Sundarban*,
SpringerBriefs in GIS, https://doi.org/10.1007/978-3-032-13981-8_7

2013; Zhang et al., 2004). According to the report of IPCC (Intergovernmental Panel for Climate Change) 2018, it has been stated that the sea level has gradually increased after 2010 (Masson-Delmotte et al., 2018). Due to the storm surges, sea level rise, coastal flooding, and low-lying intrusion, the delta, estuaries, islands, and shore areas are more vulnerable and exposed in the current situation, which also damages the coastal infrastructure. Coastal erosion creates an immense impact on socio-economic conditions in any coastal region because these zones are highly populated due to the low-lying agricultural conditions (Hinkel et al., 2010; Pirazzoli, 1996; Pye & Blott, 2006).

In this current context, erosion-accretion analysis is an acute problem for coastal zone management and sustainable eco-tourism. The coastal region is used in an unrestrained way due to the rapid enhancement of urbanization, industrial growth, fisheries, and tourism development (Chua, 1993; Nayak, 2002). In the coastal region, coastal landforms are gradually degraded and induced by several geological processes in the past (Ferraras & Pararas-Carayannis, 1988). In any coastal country, it is important to be concerned about the coastal erosion, which is also influenced by several man-made activities, and it modifies the coastal landforms (Kato & Eanhmp, 2002). In coastal areas, saltwater intrusion, mangrove loss, dune degradation, and coastal erosion are the major environmental concerns today (Mukesh, 2010), especially in India. The areas that are affected by the flooding, storm surges, sea level rise, and erosion are considered the highly susceptible hazardous coastal zone (O'Connor et al., 2009).

To balance physical and anthropogenic factors, an extensive coastal management is required for sustainable management (Espey et al., 2015; Liu et al., 2015). So, it is important to find out the various man-made and physical activities (Nichols, 1999; Sorense, 1997). After various literature surveys, it has been important to identify the vulnerable zone due to coastal erosion in the Indian Sundarbans to protect the coastal mangrove region. The erosion-accretion zonation mapping is most helpful for the coastal planner and decision-makers for sustainable development (Mujabar & Chandrasekar, 2013; Parthasarathy & Natesan, 2015). The present study aimed to identify the coastal erosion-accretion condition of the Indian part of the Sundarbans at short- and long-term scales. The study has also followed several statistical methods to assess the vulnerable condition due to erosion in this coastal region. Mangrove condition has also been analyzed using a Support Vector Machine (SVM) system to illustrate the change or modification of vegetation over 20 years in this region (2000–2020). The book has been describing the usefulness of remote sensing GIS applications for coastal zone analysis as well as the mangrove health assessment, which is also significant for further coastal zone management and decision-making studies.

7.2 Potential Use of Short-Term Shoreline Erosion-Accretion Mapping

Coastal erosion-accretion zonation mapping aimed to establish a helpful outcome for coastal decision-makers and planners to manage the coastal zone in a sustainable way. Erosion-accretion zonation mapping on a short-term scale is very crucial in the Sundarbans region due to the delta area being highly dynamic in nature and excessively vulnerable to sudden physical changes. The importance of short-term modification and erosion-accretion zonation mapping can be detailed as follows:

7.2.1 Rapid Response to Extreme Weather Events

The Sundarbans Delta and shoreline region have been witnessing several cyclones, floods, storm surges, and sea level rise. Short-term zonation mapping is most crucial to capture the immediate loss and gain after those events. Reaching out to the loss and gain condition of the region, quick decision-making could be possible to repair the embankments, protect the livelihoods, relocate the assets, and remove the saltwater from the agricultural land as early as possible.

7.2.2 Monitoring the Seasonal Variability

As the Sundarbans region is a deltaic and estuary region, the sediment transport, tidal conditions, and river channel runoff vary across the season, such as monsoon, pre-monsoon, and post-monsoon seasons. From the zonation mapping of erosion-accretion on a short-term scale, season-specific erosion hotspot areas are identified quickly, and immediate actions can be taken by the administrations.

7.2.3 Local Livelihoods Protection

The Sundarbans mangrove region is significant for various natural resources like fishing, honey collection, crab and shrimp cultivation, and agricultural activities on the delta land and estuary areas. So, the seasonal erosion-accretion zonation study in the short term will be very helpful for early warning to the fishers, honey collector to protect their livelihoods.

7.2.4 Early Warning System for Infrastructure Damage

As the Sundarbans areas are estuary and deltaic, the islands are connected with jetties, and due to storms, a high alert zone and many cyclonic areas have been established here. A lot of embankment has also been established on the river side so the local native people could be protected. Short-term, frequent zonation mapping will easily identify the most vulnerable locations, allowing infrastructure to be established before major damage occurs.

7.2.5 Mangrove Protection

The Sundarbans Delta is significant for the mangrove vegetation that also protects from several natural hazards, disasters, and cyclones. Coastal erosion is one of the major reasons for destroying the mangrove vegetation areas; on the other hand, sedimentation can regenerate the potential mangrove areas. From the short-term shoreline change or erosion-accretion zonation mapping, the erosion-prone areas are easily identified over the Indian part of the Sundarbans, and accretion areas are observed to regenerate the mangrove plantation. From the early warning system, erosion hotspots can be protected and mitigated to reduce excessive erosion, helping reduce mangrove loss.

7.2.6 Disaster Preparedness

Short-term zonation mapping can be incorporated in disaster management decision-making to improve the shelter locations, evacuation routes, and rescue process during emergencies. From shoreline modification data, coastal planners will obtain the proper erosion-accretion rate data before constructing the new infrastructure and applying protective mitigation strategies.

7.3 Potential Use of Long-Term Shoreline Erosion-Accretion Zonation

The current research work has not only detailed the short-term erosion-accretion zonation mapping but also prepared the long-term change in shoreline condition over the Sundarbans region during the 2000–2020 time period using several statistical techniques like linear regression rate (LRR) and net shoreline movement (NSM). To protect the Sundarbans evolving earth surface from natural and man-made activities, the long-term coastal erosion-accretion zonation or decadal studies play a vital role. The significance of long-term zonation work has been summarized in detail below:

7.3.1 *Identify the Determined Trend of Shoreline Modification*

Long-term change studies and zonation maps have revealed the cumulative coastline change condition, which is neglected in short-term analysis. From temporary shift to permanent loss and gain, the condition of the region has been assessed in a long-term zonation study. A decadal change can only have been assessed in long-term erosion-accretion zonation work that helps to identify the future trend of erosion and sedimentation in the delta region.

7.3.2 *Strategic LULC Planning*

From the long-term analysis safe location can be chosen for agriculture, settlement, and infrastructure development away from the hotspot of erosion. A long-term zonation map will be helpful for reaching out to the saltwater intrusion zone to reduce the loss of agricultural land. From the long-term analysis, eroded areas were avoided from development and tourism; on the other hand, accreted areas were used for further agricultural development as stable land.

7.3.3 *Climate Change and Assessment of Sea Level Rise*

From the long-term zonation mapping, sediment transport conditions have been assessed for sea level rise. On the other hand, the long-term erosion-accretion data future inundation triggering condition has been assessed under climate change scenarios. Long-term modification of the shoreline will be a potential for finding out the future trend of regression and transgression of the coastline.

7.4 Conclusion

The present author's book has implemented the shoreline modification rate and prepared the zonation map based on short-term and long-term shoreline modification rates using the DSAS statistical analysis system through remote sensing and GIS applications. The chapter tried to conclude about the potential use of the short and long-term zonation map of the Indian Sundarbans region, along with various islands. The Sundarbans is not only rich in mangrove but also has a versatile environmental nature. This region is dominated by various cyclonic hazards and disasters, so it has been crucial to identify the shoreline modification condition and the erosion-accretion trend. For any coastal circumstance, the transgression and regression of the shoreline are very crucial for sustainable development management and decision-making policies. After the extensive analysis of shoreline change rate, it can be concluded that most of the island and shore areas are found under an erosional trend that has also been triggering and susceptible to mangrove growth. This long and short-term zonation is a very strategic tool for sustainable development, climatic adaptation, and biodiversity protection by revealing the interference zone of land and water. In the case of the Sundarbans Delta region, mangrove vegetation is the primary concentration for reducing natural disasters. From the change analysis, most sediment accumulation zones are easily identified, so the mangrove can easily regenerate there.

References

Arkema, K. K., Guannel, G., Verutes, G., Wood, S. A., Guerry, A., Ruckelshaus, M., Kareiva, P., Lacayo, M., & Silver, J. M. (2013). Coastal habitats shield people and property from sea-level rise and storms. *Nature Clinical Practice Endocrinology & Metabolism, 3*, 913.

Chua, T. E. (1993). Essential elements of Integrated Coastal Zone Management. *Ocean and Coastal Management, 21*, 81–108.

Donchyts, G., Baart, F., Winsemius, H., Gorelick, N., Kwadijk, J., & van de Giesen, N. (2016). Earth's surface water change over the past 30 years. *Nature Clinical Practice Endocrinology & Metabolism, 6*, 810.

Espey, J., Swanson, E., Badiee, S., Christensen, Z., Fischer, A., Levy, M., et al. (2015) Data for development: A needs assessment for SDG monitoring and statistical capacity development. Retrieved January 22, 2018, from https://sustainabledevelopment.un.org/content/documents/2017Datafor-Development-Full-Report.pdf.

Ferraras, S. F., & Pararas-Carayannis, G. (1988). Natural and man-made coastal hazards. In *Proceedings of the 3rd International Conference on Natural and Man-Made Coastal Hazards Held in August 15–20, 1988* (pp. 37–42). Ensenada, Baja California.

Hinkel, J., Nicholls, R. J., Vafeidis, A. T., Tol, R. S. J., & Avagianou, T. (2010). Assessing risk of an adaptation to sea-level rise in the European Union: An application of diva. *Mitigation and Adaptation Strategies for Global Change, 15*, 703–719.

Kantamaneni, K., Phillips, M., Thomas, T., & Jenkins, R. (2018). Assessing coastal vulnerability: Development of a combined physical and economic index. *Ocean and Coastal Management, 158*, 164–175.

Kato, H., & Eanhmp. (2002). *Eastern Asia natural hazard mapping project* (p. 584). Explanatory Note of East Asia Geological Hazards Map, Geological Survey of Japan, AIST.

Leatherman, S. P., Zhang, K., & Douglas, B. C. (2000). Sea level rise shown to drive coastal erosion. *EOS Transactions of the American Geophysical Union, 81*, 55–57.

Lins-de-Barros, F. M. (2017). Integrated coastal vulnerability assessment: A methodology for coastal cities management integrating socioeconomic, physical and environmental dimensions—Case study of Regido dos Lagos, Rio de Janeiro, and Brazil. *Ocean and Coastal Management, 149*, 1–11.

Liu, J., Hull, V., Mooney, H., Gaskell, J., Davis, S. J., Hertel, T., Seto, K. C., Lubchenco, J., Gleick, P., Kremen, C., et al. (2015). System Integration for global sustainability. *Science, 347*, 963–972.

Masselink, G., & Hughes, M. G. (2014). *An introduction to coastal processes and geomorphology*. Routledge.

Masselink, G., & Russell, P. (2013). Impacts of climate change on coastal erosion. *MCCIP Science Review, 2013*, 71–86.

Masson-Delmotte, V., Zhai, P., Pörtner, H. -O., Roberts, D., Skea, J., Shukla, P. R., et al. (Eds.). (2018). Summary for policymakers. In: Global warming of 1.5 °C. An IPCC special report on the impacts of global warming of 1.5 °C above pre-industrial levels and related global greenhouse gas emission pathways, in the context of strengthening the global response to the threat of climate change, sustainable development, and efforts to eradicate poverty (p. 32). Geneva, Switzerland: World Meteorological Organization.

Mujabar, P. S., & Chandrasekar, N. (2013). Coastal erosion hazard and vulnerability assessment for Tamil Nadu south coast of India by using RS and GIS. *Natural Hazards, 69*, 1295–1314.

Mukesh, S. B. (2010). Coastal vulnerability, adaptation and risk assessment due to environmental change in Apodi Mossoro estuary, Northeast Brazil. *International Journal of Geomatics and Geosciences, 1*(3), 620–638.

Nayak, S. (2002). Use of satellite data in coastal mapping. *Indian Cartographer, 22*, 147–156.

Nicholls, R. J., Wong, P. P., Burkett, V., Codignotto, J., Hay, J., McLean, R., et al. (2007). *Coastal systems and low-lying areas*.

Nichols, K. (1999). Coming to terms with "integrated coastal management": Problems of meaning and method in a new arena of resource regulation. *The Professional Geographer, 51*(3), 388–399.

O'Connor, M. C., Lymbery, G., Cooper, J. A. G., Gault, J., & Mckenna, J. (2009). Practice versus policy-led coastal defence management. *Marine Policy, 33*, 923–929.

Parthasarathy, A., & Natesan, U. (2015). Coastal vulnerability assessment: A case study on erosion and coastal change along Tuticorin, Gulf Mannar. *Natural Hazards, 75*, 1713–1729.

Pirazzoli, P. A. (1996). *Sea-level changes: The last 20,000 years*. Chichester: Wiley.

Pye, K., & Blott, S. J. (2006). Coastal processes and morphological change in the Dunwich—Sizewell area, Suffolk, UK. *Journal of Coastal Research, 22*, 453–473.

Shepard, C. C., Agostini, V. N., Gilmer, B., Allen, T., Stone, J., Brooks, W., & Beck, M. W. (2011). Assessing future risk: Quantifying the effects of sea level rise on storm surge risk for the southern shores of Long Island, New York. *Natural Hazards, 60*, 727–745.

Sorensen, J. (1997). National and international efforts at integrated coastal management: Definitions, achievements and lessons. *Coastal Management, 25*(1), 3–41.

Zhang, K., Douglas, B. C., & Leatherman, S. P. (2004). Global warming and coastal erosion. *Climatic Change, 64*, 41–58.

MIX
Papier aus verantwortungsvollen Quellen
Paper from responsible sources
FSC® C105338

If you have any concerns about our products, you can contact us on
ProductSafety@springernature.com

In case Publisher is established outside the EU, the EU authorized representative is:
Springer Nature Customer Service Center GmbH
Europaplatz 3, 69115 Heidelberg, Germany

Printed by Libri Plureos GmbH
in Hamburg, Germany